Contributions from Science Education Research

Volume 7

Contributions from Science Education Research is the international, multidisciplinary book series of the European Science Education Research Association (ESERA). The aim of the series is to synthesize, for the benefit of the scholarly community, the findings of high quality, theoretically-framed research in the domain of science education as well as comprehensive explorations of specific methodological strands in science education research. The series aims to publish books that are innovative in attempting to forge new ways of representing emergent knowledge in the field. The series includes edited collections of chapters, monographs and handbooks that are evaluated on the basis of originality, scientific rigor and significance for science education research. The book series is intended to focus mainly on work carried out in Europe. However, contributions from researchers affiliated with non-European institutions and non-members of the European Science Education Research Association are welcomed. The series is designed to appeal to a wide audience of researchers and post-graduate students in science education. Book proposals for this series may be submitted to the Publishing Editor: Claudia Acuna E-mail: Claudia.Acuna@springer.com

More information about this series at http://www.springer.com/series/11180

Laurence Viennot • Nicolas Décamp

Developing Critical Thinking in Physics

The Apprenticeship of Critique

Springer

Laurence Viennot
Matter and Complex Systems UMR 7057
University of Paris
Paris, France

Nicolas Décamp
Laboratoire de Didactique André
Revuz EA 4434
University of Paris
Paris, France

Translation from the French language edition: L'APPRENTISSAGE DE LA CRITIQUE DÉVELOPPER L'ANALYSE CRITIQUE EN PHYSIQUE by Laurence Viennot and Nicolas Décamp, © EDP Sciences and UGA Editions, 2019. Original Publication ISBN 978-2-7598-2355-0 and 978-2-37747-091-4. All rights reserved.

ISSN 2213-3623 ISSN 2213-3631 (electronic)
Contributions from Science Education Research
ISBN 978-3-030-43772-5 ISBN 978-3-030-43773-2 (eBook)
https://doi.org/10.1007/978-3-030-43773-2

This Springer imprint is published by the registered company Springer Nature Switzerland AG.
The registered company address is: Gewerbestrasse 11, 6330 Cham, Switzerland

Contents

Chapter 1
Why Critique? Why Physics?

Imagine you read the following text about a freefall parachute jump from high altitude in a popular magazine:

> To achieve this exploit, he (the hero) will be equipped with a pressurised suit similar to those used by astronauts, but adapted to resist extremely low temperatures below 110 Kelvin degrees and equipped with a parachute. He will reach an altitude of 40,000 meters in about three hours, aboard a capsule, which is also pressurised, attached to a helium balloon. The duration of the jump is about six minutes and twenty five seconds. In the absence of an atmosphere, Fournier will pass the speed of sound (1067 kilometers/hour) about thirty seconds after his departure in a vertical position.

This text was proposed by official education authorities in France (Ministère de l'Éducation Nationale 2010) to be used to foster critical thinking in grade 10 students, and to teach them how to "extract information" from this type of resource. Teachers were advised to ask their pupils questions like: Why jump from such a high altitude? Why a helium balloon? And so on. Imagine now that you want to decide which questions to pose to pupils at that level to help them draw the maximum benefit from this text. Stop reading a few seconds and think: which questions would I ask? We put this question to several groups of physics teachers: PhD students already teaching physics at university (N = 23), physics teachers at university (N = 10), pre-service physics teachers already teaching at secondary school level (N = 38). Various suggestions were made regarding questions to be posed to students about this text: Nothing in these questions called attention on the strange assertion that at the starting point of the fall there was no longer air ("in the absence of an atmosphere") but there was a balloon.

It is not unusual to hear a teenager say that all this is due to the fact that a helium balloon necessarily rises. But the teachers consulted were not beginners in physics, far less the group of experts in charge of drafting official documents. They all knew Archimedes' principle, that is, how to calculate the force exerted by a fluid, in a gravitational field, on a body immersed in it. Despite this, only one teacher in the groups indicated that the lack of atmosphere was worth questioning.

L. Viennot, N. Décamp, *Developing Critical Thinking in Physics*,
Contributions from Science Education Research 7,
https://doi.org/10.1007/978-3-030-43773-2_1

This kind of critical passivity of experts may seem surprising but it is not an isolated example, as we will show. It raises a series of questions all linked to what is often termed “critical thinking”.

1.1 Developing Critical Thinking: A Tradition and an Urgency

Critical thinking is nowadays a widely advocated competence. In France particularly, secondary scientific education is strongly focused on exploiting resources designed for popular audiences, with an explicit emphasis on critical judgment. The syllabus for grade 12 scientific students (Bulletin Officiel de l’Education Nationale 2011) repeatedly advocates (18 occurrences) the activity of “extracting and using information” from various documents. It stipulates that the activities proposed to students concerning the competence of “extracting”, and the knowledge they acquire, should lead them to raise critical questions about the scientific value of information sources, about the relevance of taking such questions into account, and about how to choose what to retain from a set of resources where there is often superfluous and sometimes inaccurate information, and where it is necessary to distinguish objective and rational knowledge from opinion and beliefs. More than twenty years ago, Millar led a project – Twenty First Century Science – which included an emphasis on helping students to access and interpret texts about science written for the general public (Millar 1996; 21st Century Science Project Team 2003). Millar (2006) draws on the views of Norris and Phillips (2003) who argue that “Science (…) could not exist as an oral tradition; texts are essential, not optional. They are a constitutive feature of science – just as empirical data collection is. An understanding of science therefore requires the ability to read texts” (p. 1502).

To our knowledge, there are no explicit arguments against this teaching objective in the academic sphere. But what about how to achieve it? How can we develop this valuable ability to evaluate thoughtfully the arguments of others and how can we support the development of greater competence in this regard?

First, what is critical thinking? Looking at research literature, we find very large definitions. Willingham (2007), for instance, writes that critical thinking concerns “critical reasoning, decision making and problem solving” (p. 11). Ennis (1989) says that critical thinking is “reasonable reflective thinking focused on deciding what to believe or do” (p. 4). McPeck (1981) sees it as “the correct assessing of statements” and “the propensity and skill to engage in an activity with reflective skepticism” (p. 152). Still wider is the perspective recently developed by Noddings and Brooks (2016): Proper critical thinking would be first and foremost searching for meaning. Looking beyond the differences in these various standpoints, what we retain is a positive perspective on critical thinking, that is, we see critical thinking as a path toward better comprehension rather than the art of defeating an opponent by argument.

That said, we adopt in this book a more limited approach to critical thinking than is suggested by these definitions. In particular, we do not consider "controversial" or "socio-scientific" issues. Much has been written about these, particularly concerning the status of experts, their political or economic independence (Jiménez-Aleixandre and Puig 2009) and the mechanisms for spreading rumours. On these issues, important work is being carried out by groups such as the Rationalist Union, founded in 1930 through the efforts of Paul Langevin, or more recently CorteX, a group of scientists who are committed to promoting critical analysis and rational doubt (for example, Bronner 2013; Gauvrit 2007; see also Ellenberg 2015). In France, the term "zetetics" (or "theory of doubt") has been used to designate this intellectual enterprise. Monvoisin (2007) suggests that *(...)* "zetetics is nothing other than the scientific method, but applied to fields of knowledge which carry such an affective load that it requires the integration of intellectual impasses and cognitive biases related to belief, adherence or commitment" (p. 152, in French). Among the questions addressed in these investigations are the reasons why so many people prefer to think that the Earth is flat, and that the photos showing the opposite are fake. All these questions are essential, but not our focus here.

In this terminology, this book is about zetetics applied to the field of physics, excluding socio-scientific components. The choice to exclude these does not mean that they are less important concerns, far from it. It is fortunate that these major issues have been recognized and addressed for several decades by scientists who are engaged in their analysis and in constructing warning systems for them. An essential component of their work is the identification of psycho-cognitive aspects, such as the compelling desire for an explanation when none is yet available within accepted science, a desire that leads to prefer a non-rational explanation to no explanation at all; or the "confirmation bias" which consists in ascribing more value to arguments supporting what we already think than to the others (Henderson et al. 2015); or the conspiracy theories that accompany certain attitudes of systematic doubt – for instance about the reasons why the World Trade Center towers collapsed on 11th September 2001.

This book targets a context that is in principle less complex, that of texts commonly used in physics education. So, the phenomena that we wish to illustrate do not focus on aspects with the kind of "emotional load" raised by so-called socio-scientific issues. For instance, the reason why a helium balloon holds in the air does not raise the same type of difficulty as the influence of humans on the climate.

That said, why focus on a particular domain of knowledge, and why on physics?

1.2 Why a Particular Domain of Knowledge?

As regards the first question, our approach is based on the view that it is valuable but insufficient to know general principles of critical text analysis, such as source control or knowledge of the main cognitive biases. Historically, this assumption of insufficiency has been the subject of lengthy debate. Some researchers (including

Ennis 1996; Halpern 1998; Kuhn 1999; Davies 2013) have argued that critical thinking competencies are cross-cutting and can be applied to many areas, while others (for example, McPeck 1981, 1990; Barrow 1991; Bailin 2002; Willingham 2007) have highlighted the specificity of the criteria for critical thinking in different areas. Several authors have carried out reviews of the large number of studies published on this topic (Abrami et al. 2008). On this basis, it has been suggested that explicit initiation into the main principles of critical thinking should be combined with education in critical analysis in a given field (Davies 2013; Tibebu Tiruneh et al. 2017).

With this in mind, we wish to contribute mainly to this second component of critical education, because the need for working tools in the physical sciences still seems to us to be insufficiently met. Indeed, if we think of beginning teachers of physics, many of them are already aware of the difference between science and pseudo-science and of the need to verify sources. Despite their relative inexperience, they already know how difficult it is to convince some students that the Earth is not flat. But still they are stunned when they are led to realize that they themselves can fail to be critically aware in the familiar terrain of ordinary physics teaching.

Alongside this content-specific focus, we have, however, a wider ambition, to illustrate by means of physics some phenomena that are likely also to intervene in other fields of knowledge. Like La Fontaine's fables, the situations evoked can be used as analogical patterns for a set of phenomena observable in other contexts. We will therefore encounter possible resonances from one discipline to another, from one context to another. For instance, the text about the world record of free fall cited above can be seen as equating "small" and "zero". At an altitude of 40 km, air pressure is only a few thousands of its value at sea level. Can we "neglect" this value and claim that, at this altitude "there is no atmosphere" or, more precisely, that the value of atmospheric pressure is zero? In fact, it all depends what phenomenon we are analysing. As regards the fall of the skydiver, it is quite acceptable to consider this value negligible and conclude that, at this altitude, he is in free fall. But concerning the presence of a helium balloon, equating small and zero leads to absurdity, because the non-zero value of air pressure is essential for the support of the balloon (thanks to the very large volume of this balloon, it is also sufficient). This kind of discussion – should we neglect a "small" quantity, with respect to what and in which context? – can be useful in many other disciplines, for instance life sciences or economics.

1.3 Why Physics?

So why do we focus on physics?

We might simply say that physics is an area, like many others, where critical judgment is of crucial importance. But in fact, physics lends itself particularly well to education in critical analysis. A favourable factor is the existence of well-defined and measurable quantities. These also exist in other fields, but in physics they are the rule. This permits a prediction that arises from reasoning to be tested quantitatively. Moreover, physics is a very structured science where a few laws make it possible to account for many situations. This is the substance of its remarkable successes, and also why it is periodically forced to undergo substantial restructuring – or even "revolutions" (Kuhn 1962). In brief, physics imposes very strong constraints on reasoning (for an outline of our epistemological view on reasoning in physics, see Appendix A), whether in terms of explanation or prediction. These constraints must all be met, first and foremost the compatibility between what we say and our observations and measurements. As is often said, in physics it is necessary to "put together" many things, and an "ad hoc" explanation is not appropriate. Hence the interest of using physics as a privileged field for the development of critical analysis, even if thinking like a physicist is not the only way to be rational, far from it.

In addition, much research material is now available to shed light on common patterns of thinking as they manifest themselves in physics (Duit 2009; Viennot 2001), for instance, the common tendency to think of a "force" as a property of an individual or an object that explains (is proportional to) its motion (its speed). This research makes it possible to anticipate, or at least to easily identify, certain characteristics of reasoning that are not consistent with the physics of experts and that, when observed, merit critical analysis.

1.4 Educating Teachers on Critical Analysis

But our aim is not primarily to explore further these common patterns of thinking, it is above all to see how to achieve a real development of critical analysis in physics. The elements we propose focus on the preparation of beginning teachers which seems to us a logical beginning as we must rely on them to educate their students.

When undertaking such a project, the foreseeable obstacles cannot be ignored. Generally speaking, the positions of influential researchers in cognitive psychology are rather discouraging in this regard. "Why is it so difficult to teach critical thinking?" asks Willingham (2007); in exploring this question, he stresses in particular that "The difficulty lies not in thinking critically, but in recognizing when to do so, and in knowing enough to do so successfully" (p. 15). Kahneman (2012), for his part, considers "Fast thinking or System 1 – a set of automatic and often unconscious processes that underlie intuitive thinking"(p. 13)- as a major obstacle to the exercise of criticism. Concerning education to a better control of System 1, he writes that "little can be achieved without a considerable investment of effort. As I know from experience, System 1 is not readily educable" (p. 417). Among his conclusions, he comments: "I have improved only in my ability to recognize situations in which errors are likely (…)" (p. 417). Note, however, that developing this ability already constitutes a valuable step and we share this goal. However, in physics for instance, many of the observed difficulties require a detailed analysis that goes beyond general psycho-cognitive effects, relevant though these may be. In other words, it is not simple to anticipate the precise forms of inappropriate ways of reasoning in a particular domain, even with a good knowledge of System 1. This is not far from Kahneman's own conclusion: "System 1 registers the cognitive ease with which it processes information, but it does not generate a warning signal when it becomes unreliable. (..) There is no simple way for System 2 to distinguish between a skilled and a heuristic response. Its only recourse is to slow down and attempt to construct an answer on its own, which it is reluctant to do because it is indolent" (p. 416). In Kahneman's terms, a crucial question is therefore whether (and, if so, how) an individual desiring to solve a problem or take a decision can overcome their cognitive ease or indolence. His answer is very pessimistic.

Houdé (2014) for his part raises the possibility of inhibiting the impulsive and uncontrolled thought of System 1, thus giving a chance for critical analysis education. According to him, "Only the learning of inhibition is effective, which activates a region of the brain (right ventromedial prefrontal area) involving close relationships between emotion, self-feeling and reasoning: here emotion is linked to the feeling of having made a mistake, specifically triggered by learning to inhibit perceptual bias.(...) These brain imaging results therefore suggest that emotion can help reasoning, contrary to the idea introduced by Descartes (and still implicit in Piaget and even Kahneman), of a necessary opposition of reason and emotion" (p. 88, in French).

This view echoes positions that are widely represented in the research literature about metacognition (Flavell 1987): "Metacognition refers to learners' views and beliefs about learning and to the active regulation of their learning processes" (p. 25). This would also apply to critical thinking. It is also worth noting that expressing dissatisfaction with an explanation involves certain emotions, particularly because asking questions to achieve a deeper understanding is a demanding approach. Such an attitude implies a certain degree of self-esteem and a search for intellectual satisfaction (Bandura 2001). At least in this sense, we agree with Houdé's assertion that emotion can help reasoning, while keeping in mind that it can also hinder rationality. In any event, these very general positions should be complemented by more detailed studies on critical education in a particular discipline, such as physics. Studies of this kind are a real need.

In terms of the obstacles to be anticipated, how should we locate physics? Is it simple to develop a more critical mindset among apprentice physics teachers? Are experienced physics teachers, we might ask, experts in critical analysis? Our answer would be, yes and no.

Yes, a priori, because they know many things and therefore have important means of identifying erroneous predictions, violations of laws or flawed experimental procedures. As a result, many mistakes are obvious to them. They also have at their disposal a battery of procedures for detecting claims that are certainly inaccurate. For example, the examination of orders of magnitude, which will make them reject a mass of a few tons for a spider, or an awareness of prohibited values, such as a probability greater than 1 or a speed higher than that of light in vacuum. Or dimensional analysis, which prohibits the declaration of equality between inhomogeneous expressions (expressions whose numerical values would be changed by a different factor when changing units). In the teaching arsenal there is also the consideration of 'limiting cases' to identify implausible values that would be taken by quantities that depend on a variable when it tends, for example, towards zero or infinity.

But also *No*, because these means remain insufficient, as we will show. The meshes of this net allow highly questionable lines of reasoning to pass. In fact, we as teachers are particularly exposed to the risk of lack of critical judgment because of the influence of "teaching rituals" (Viennot 2006). Some experimental demonstrations or lines of reasoning which are embedded in the most common teaching practices have a fallacious aspect. Think for instance of hot air balloons holding in the air commonly declared to be in an isobaric situation, on the grounds that the envelope is open: How could the envelope be inflated in such a case, and how could the balloon hold in an isobaric atmosphere? Such rituals may hamper our critical capacity in the field of our discipline. It is also necessary to acknowledge the intrinsic difficulty of conducting valid reasoning in physics: who can claim always to free themselves from the difficulties generated, in particular, by the multiplicity of variables involved in most physical situations or by the difficulties of interpreting experiments (Vilches and Gil-Perez 2012) – points that we will examine later in more detail. Also, how are we to avoid allowing the statement "a model is never perfect" to become "anything – even an isobaric hot air balloon – may be envisaged as a preliminary model". We therefore start from the premise that it is highly justified to focus on teacher education, starting with those who are new to critical analysis.

1.5 Structure of the Book

This book will first characterize the types of flaws that should lead a reader to reject or question texts relating to physics (Chap. 2). The term "text" is to be taken here in the very broad sense of an identifiable source of information: a succession of sentences accompanied or not by lines of calculation, images or diagrams, all within an academic or popularization framework. We will also detail the types of text that, without making obvious mistakes, are nonetheless likely to suggest them by favouring critical passivity in readers and therefore need to be looked at particularly critically (Chap. 3). We will also point out how the interest of critical analysis goes well beyond the mere hunt for statements, suggestions or faulty reasoning from the point of view of accepted physics. We will point out what critical analysis can bring to the reading of a text in order to appreciate its content and assess its scope, in particular the possibilities of generalising from it (Chap. 4).

Elements of information gathered from novice teachers, either in individual interviews or group discussions (Viennot and Décamp 2018), will then be discussed. A crucial question arises in this regard: what are the links, in this population, between the development of a critical attitude and the development of comprehension of the content in question? In other words, can the expression of critical judgment be developed in apprentice physics teachers without the support of a conceptually sound foundation? This question becomes particularly acute at the end of two decades when, especially in France, conceptual structuring has *de facto* been overshadowed by education objectives centered on content-independent competences. It is in line with the broader question of whether or not the manifestation of a critical attitude depends on the content being examined. This question led us to conduct a series of investigations which we will briefly report, the results of which suggest an entangled development of critical attitude and conceptual understanding. These investigations, which are still exploratory, allow a first characterization of the intellectual dynamics of novice teachers on this twofold level: conceptual and critical (Chaps. 5 and 6).

Note that, in carrying out and reporting this work, we do not intend to designate categories of people who are more or less critical, because in all of us there is an existing ability to formulate critical comments, even if their expression may be delayed. We prefer to see a critical attitude (as contrasted with critical passivity) as indicating the activation (as opposed to inhibition) of a widely shared "potential for critique". Consistently, our point is to identify the types of difficulty which are likely to hinder a critical analysis of texts.

Finally, avenues will be proposed for the development of expertise in critical education, in physics and elsewhere, at various school or academic levels (Chaps. 7 and 8).

Throughout the book, our aim is not to point out errors gratuitously but to develop a vision of critical analysis as a stepping stone to advance in understanding. What this book aims to highlight are the difficulties inherent in physics, even at a so-called "elementary" level, their very general character and how some tools can help us to combine successfully criticism and better understanding.

References

21st Century Science Project Team. (2003). 21st century science—A new flexible model for GCSE science. *School Science Review, 85*(310), 27–34.

Abrami, P. C., Bernard, R. M., Borokhovski, E., Wade, A., Surkes, M. A., Tamim, R., & Zhang, D. (2008). Instructional interventions affecting critical thinking skills and dispositions: A stage 1 meta-analysis. *Review of Educational Research, 78*(4), 1102–1134. https://doi.org/10.3102/0034654308326084.

Bailin, S. (2002). Critical thinking and science education. *Science & Education, 11*, 361. https://doi.org/10.1023/A:1016042608621.

Bandura, A. (2001). Social cognitive theory: an agentic perspective. *Annual Review of Psychology, 52*, 1–26.

Barrow, R. (1991). The generic fallacy. *Educational Philosophy and Theory, 23*(1), 7–17.

Bulletin Officiel de l'Education Nationale n°8, 13th October 2011.

Bronner, G. (2013). *La Démocratie des crédules*. Paris: PUF.

Davies, M. (2013). Critical thinking and the disciplines reconsidered. *Higher Education Research & Development, 32*(4), 529–544. https://doi.org/10.1080/07294360.2012.697878.

Duit, R. (2009). *Bibliography STCSE, Students' and Teachers' Conceptions and Science Education*. http://www.ipn.uni-kiel.de/aktuell/stcse/stcse.html

Ellenberg, J. (2015). *How not to be wrong*. London: Penguin.

Ennis, R. H. (1989). Critical thinking and subject specificity: Clarification and needed research. *Educational Researcher, 18*(3), 4–70.

Ennis, R. H. (1996). *Critical thinking*. New York: Prentice Hall
Halpern, D. F. (1998). Teaching critical thinking for transfer across domains. *American Psychologist, 53*(4), 449–455. https://doi.org/10.1037//0003-066X.53.4.449.
Flavell, J. H. (1987). Speculations about the nature and development of metacognition. In F. E. Weinert & R. Kluwe (Eds.), *Metacognition, motivation, and understanding* (pp. 21–29). Hillsdale: L. Erlbaum Associates.
Gauvrit, N. (2007). *Statistiques méfiez-vous!* Paris: Ellipse.
Henderson, J. B., MacPherson, A., Osborne, J., & Wild, A. (2015). Beyond construction: Five arguments for the role and value of critique in learning science. *International Journal of Science Education, 37*(10), 1668–1697. https://doi.org/10.1080/09500693.2015.1043598.
Houdé, O. (2014). *Le raisonnement*. Paris: PUF.
Jiménez-Aleixandre, M. P., & Puig, B. (2009). Argumentation, evidence evaluation and critical thinking. In B. J. Fraser, K. Tobin, & C. McRobbie (Eds.), *Second international handbook of science education* (pp. 1001–1015). Dordrecht: Springer. https://doi.org/10.1007/978-1-4020-9041-7.
Kahneman, D. (2012). *Thinking fast and slow*. London: Penguin books.
Kuhn, T. S. (1962). *The structure of scientific revolutions*. Chicago: University of Chicago press.
Kuhn, D. A. (1999). Developmental model of critical thinking. *Educational Researcher, 28*(2), 16–25. https://doi.org/10.2307/1177186.
McPeck, J. (1981). *Critical thinking and education*. New York: St Martin's Press.
McPeck, J. E. (1990). Critical thinking and subject specificity: A reply to Ennis. *Educational Researcher, 19*(4), 10–12.
Millar, R. (1996). Towards a science curriculum for public understanding. *School Science Review, 77*(280), 7–18.
Millar, R. (2006). Twenty first century science: Insights from the design and implementation of a scientific literacy approach in school science. *International Journal of Science Education, 28*(13), 1499–1521.
Ministère de l'Éducation Nationale. (2010). *Ressources pour la classe de seconde générale et technologique*. Available from: http://cache.media.eduscol.education.fr/file/SPC/92/8/LyceeGT_Ressources_2_Commun_SPC_Sport_149928.pdf
Monvoisin, R. (2007). *Pour une didactique de l'esprit critique. Zététique & utilisation des interstices pseudoscientifiques dans les médias*. Grenoble: Université Grenoble 1 Joseph Fourier.
Noddings, N., & Brooks, L. (2016). *Teaching controversial issues: The case for critical thinking and moral commitment in the classroom*. New York: Teachers College Press.
Norris, S., & Phillips, L. (2003). How literacy in its fundamental sense is central to scientific literacy. *Science Education, 87*(2), 224–240.
Tibebu Tiruneh, D., De Cock, M., & Elen, J. (2017). Designing learning environments for critical thinking: Examining effective instructional approaches. *International Journal of Science and Mathematics Education, 16*, 1065–1089. https://doi.org/10.1007/s10763-017-9829-z.
Viennot, L. (2001). *Reasoning in physics the part of common sense*. Dordrecht: Kluwer. (now Springer).
Viennot, L. (2006). Teaching rituals and students' intellectual satisfaction. *Physics Education, 41*, 400–408. http://stacks.iop.org/0031-9120/41/400.
Viennot, L., & Décamp, N. (2018). Activation of a critical attitude in prospective teachers: From research investigations to guidelines for teacher education. *Physical Review Physics Education Research., 14*, 010133. https://doi.org/10.1103/PhysRevPhysEducRes.14.010133.
Vilches, A., & Gil-Perez, D. (2012). The supremacy of the constructivist approach in the field of physics education: myths and real challenges. *Tréma, 38*, 87–106.
Willingham, D. T. (2007). Critical thinking why is it so hard to teach? *American Educator*, 1–19. Retrieved from http://www.aft.org/sites/default/files/periodicals/Crit_Thinking.pdf

Chapter 2
Main Reasons to Call a Text into Question

Let us start with simple and rather commonplace ideas. From the perspective of education, our aim is to document these ideas by remarks and examples that, in contrast, will not seem self-evident. Below is a list of the main pitfalls that should be avoided, in principle, when constructing an explanation.

2.1 Internal Contradiction

Even a child might say: it's better not to make a statement along with another that contradicts it without any further comment. That seems so obvious that we can hardly imagine finding such a situation in a text dealing with science. Yet, the two examples below show that it can happen (Viennot 2013).

The first concerns a DVD aimed at promoting an experimental teaching sequence at lower secondary level (grade 7). The topic was thermal transfers with several materials. The students have to think about how to protect against cold when situated on Mars. One of them suggests using a survival blanket, which gets unanimous approval from the teacher and the students. In fact, current survival blankets are made of Mylar, a very thin material (about 10 μm thick) with a very reflecting side, which is also a property of aluminium. So aluminium is declared a good candidate to solve the problem at hand. After brainstorming about other candidate materials, glass, wood, aluminium and other materials are put to the test by a very simple experiment. Hot water is poured into a series of small identical containers, each covered by a small plate made of each of the materials to be tested, and ice cubes are put on top of each plate. Which ice cube will melt first? The one placed on aluminium! The teacher's conclusion is that "With aluminium, you cannot protect against cold". No link to the first student's remark is observable on the DVD. Two contradictory statements were therefore successively presented as valid, with no further discussion: something and its opposite.

L. Viennot, N. Décamp, *Developing Critical Thinking in Physics*, Contributions from Science Education Research 7,
https://doi.org/10.1007/978-3-030-43773-2_2

In a teacher education session, this example can be used to support a kind of existence theorem: Yes, this situation can occur. But it is not enough to detect a problem; it is better to try to understand, or at least hypothesize, the intellectual process underlying this absence of critical awereness in this instance. If thermal transfer is considered to be solely related to conduction, aluminium undoubtedly offers poor protection. Such an obvious idea masks the fact that highly reflective blankets can be useful to protect against the cold. In fact, several types of energy transfer are involved (conductive, convective and radiative) so that the performance of an experiment on only one of them – conductive transfer - does not permit the general conclusion formulated by the teacher (Viennot and Décamp 2016). This example also highlights a crucial point regarding criticism of physics explanations: what happens when a phenomenon or variable that may be relevant in the situation under study – here radiative transfer - is overlooked.

Another example is provided by a popularization text for secondary school teachers about capillary rise in vertical tubes that dip into the water in a container (Viennot 2013). A figure shows three tubes of different diameters: the thinner the tube, the higher the water rises. The account, however, says that the pressure under the meniscus is greater, and also smaller, than that of the atmosphere, depending on whether one reads the text or the figure caption.

The text states:

> If we immerse a glass tube vertically in water, we observe that the water rises a little up the walls, forming a meniscus which is convex downwards. From Laplace's law (...), this means that the pressure just below the meniscus is greater than that of the atmosphere just above. It is this imbalance that explains the rise (...). The smaller the tube diameter, the greater the curvature of the liquid surface and the greater the pressure difference between the two sides of the surface - and the greater the rise in the tube (...).

The caption also stipulates that the smaller the internal diameter, the larger the rise, but provides the following explanation:

> (…) the smaller the diameter of the tube (…), the larger the meniscus curvature, therefore, given Laplace –Young's law, the smaller the pressure of the water in the tube.

It is true that the topic of capillary rise is very complex and most of us have no idea of how to explain it. The targeted readers were in this case. We will give more details on capillary rise further in the book but the point here is just to illustrate that cases of internal contradictions do exist and can be encountered by curious people in search for information about a topic they do not know at all. In fact, neither the pressure under the meniscus nor the curvature of this meniscus can account for the phenomenon. As we will see later (in chapters 4 and 6), it is necessary to take into account the attractive forces between the glass molecules and the water molecules near the meniscus, and the pressure of the water molecules along the wall (Das et al. 2011; Marchand et al. 2011). The will to simplify, along with a causal interpretation of Laplace's formula ($\Delta P = \gamma \kappa$, relating the pressure difference ΔP between each side of the meniscus, its curvature κ, and the surface tension γ), then appear to become traps into which the author fell.

This example contributes to validating the existence theorem discussed above, that is, contradictory statements can indeed be found in the same text, even though it is written by professional scientists with the intention of coherence.

2.2 Direct Contradiction of a Law

We do not often witness such blatant contradictions as those discussed in the previous section. Yet, from the point of view of an education to critical analysis, they provide an opportunity to exercise critical competence at a fundamental level and it is not without interest that they clearly occur. These aspects (indisputably arising, relatively rare) are also present in certain examples involving an explicit contradiction between an element present in a text and a law of expert physics.

Let us take the example of Newton's third law. In a level 10 book in France we read:

> The brick exerts on a table a force $\vec{F}_{BonT}$ which is clearly equal to its weight. The table exerts a force on the brick $\vec{F}_{TonB}$ called reaction $\vec{R}$ of the table on the brick. It is logical to think that these two forces have same directions, same values and opposite directions. (…) Note : The action is equal to the reaction as long as the table is strong enough. If the weight of the book P is too large, the table will no longer be able to ensure equality $R = P$ and will collapse.

In this excerpt, the sentence "Action is equal to reaction as long as the table is strong enough" explicitly contradicts the third law, the scope of which is not limited to equilibrium. At any moment, even in case of collapse, this law remains valid in Newtonian dynamics. At equilibrium, the common value of the two forces involved in the interaction (here called "action" and "reaction") is equal to the weight of the brick. In case of collapse, this common value is zero and therefore the action of the brick on the table is no longer equal to its weight.

We can suggest an analysis of the origin of the error (Viennot 1982; Menigaux 1986; Brasquet 1999). There is confusion, or at least a failure to distinguish clearly, between Newton's second and third laws. The text argues that there is a balance between two reciprocal forces - which in fact are not exerted on the same object – whilst also suggesting that the action of the brick on the support is always equal to its weight (see Figs. 2.1 and 2.2).

But whatever the origin of this difficulty, critical judgment should be exercised with regard to the statement in question, or the notion of a law (of course accompanied by its validity domain) would be meaningless.

2.3 Indirect Contradiction of a Law

In most cases, we do not encounter contradictions as blatant as those illustrated above. But we cannot conciliate several elements of accepted physics and the text under consideration.

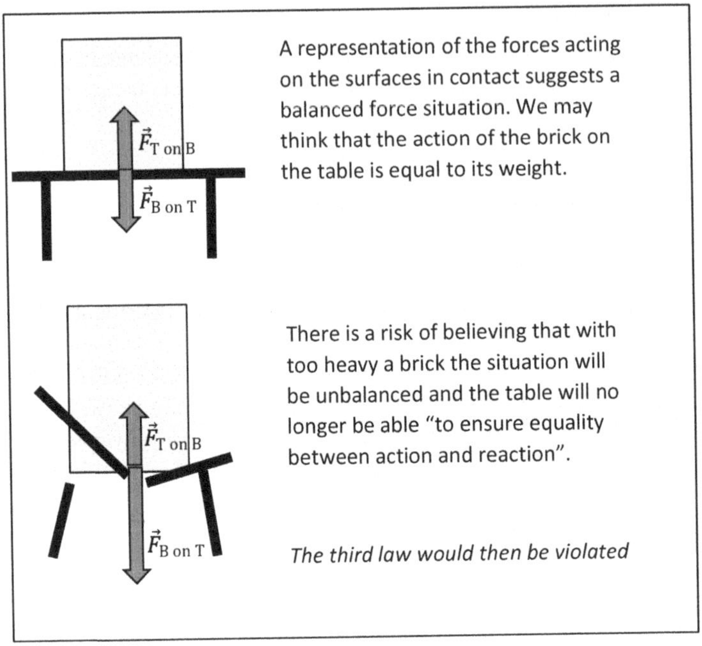

Fig. 2.1 Risks of error when analysing the situation of a brick on a table

Thus in elementary mechanics, an exercise from a Grade 12 textbook (Belin 2012) begins like this:

> A bicycle (treated as a point mass) is in uniform rectilinear motion with velocity $\vec{v} = v\overrightarrow{u_x}$. The ambient air exerts a friction force $\vec{f_f} = -\lambda \vec{v}$ on it where λ is a positive constant; the friction force exerted by the ground can be neglected. The force $\vec{F}$ exerted by the cyclist on the bicycle is in the same direction as $\vec{v}$. Express its value (…) (p. 99)

An image similar to Fig. 2.3 1 is provided with the exercise text. It is very likely that the purpose of this exercise is to get students to use a diagram like the one in Fig. 2.4.

The text says that the bicycle experiences a backward force $\overrightarrow{f_f}$, due to the ambient air, and a forward force of same direction, exerted by the cyclist. These two forces must have same value to account for the constant velocity. With this diagram, everything seems to run smoothly, at least from the point of view of Newton's second law.

Looking now to the third law, the reciprocal force (the "reaction") to the force of the air on the bicycle is directed forward, which does not raise any particular difficulty. But the reciprocal force to the so-called "force exerted by the cyclist on the bicycle" takes a surprising form, as Fig. 2.5 shows. When we try to "put together" the two laws and the text data, the cyclist falls off his bike. It is clear that something is wrong. To avoid this kind of conceptual disorder, the critical analysis will have to take into account the friction interaction between the bicycle and the ground, and give up the effort to use an internal interaction to justify the movement of the bicycle-cyclist system (Fig. 2.6).

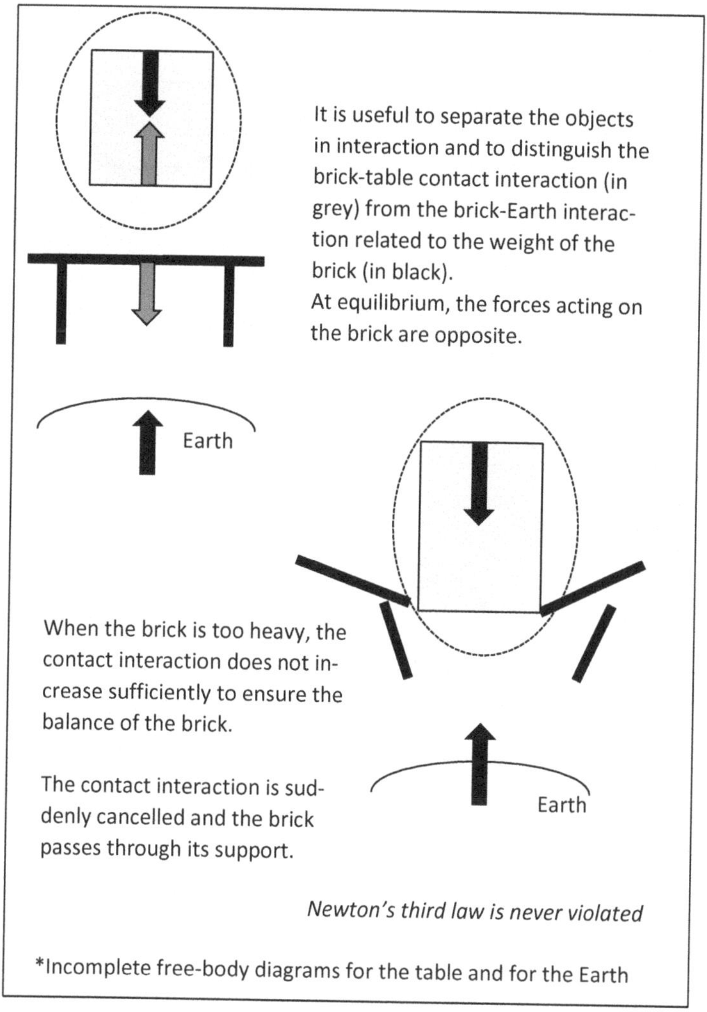

Fig. 2.2 Clarifying the situation of a brick on a table

Fig. 2.3 A cyclist at constant velocity

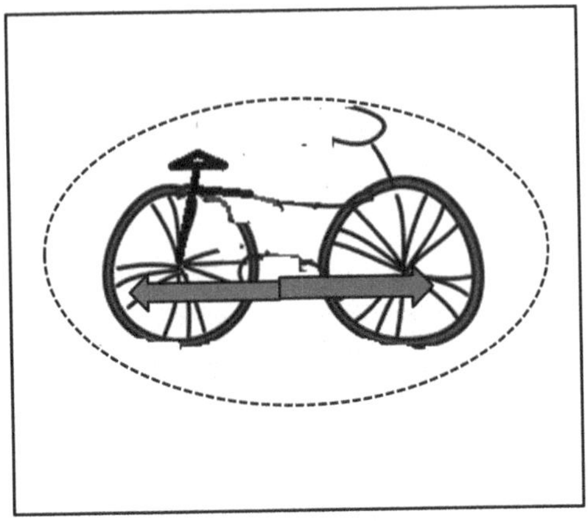

Fig. 2.4 The diagram that appears to be expected from students about the situation shown in Fig. 2.3. Backward force: friction force of air on bicycle; Forward force: force "exerted by the cyclist on the bicycle"

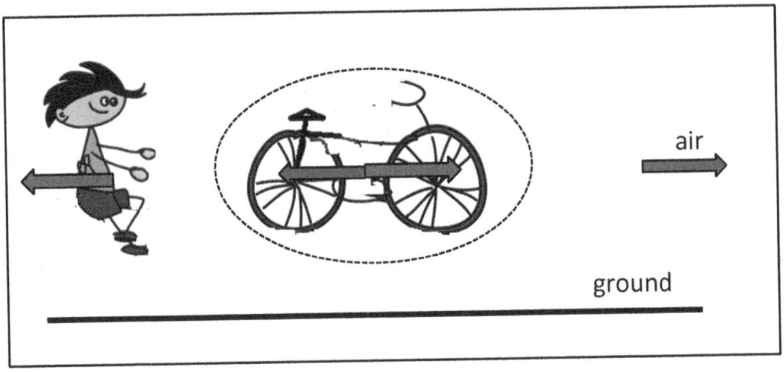

Fig. 2.5 Reciprocal forces for those shown in Fig. 2.4: the one shown on the cyclist would eject him from his bike

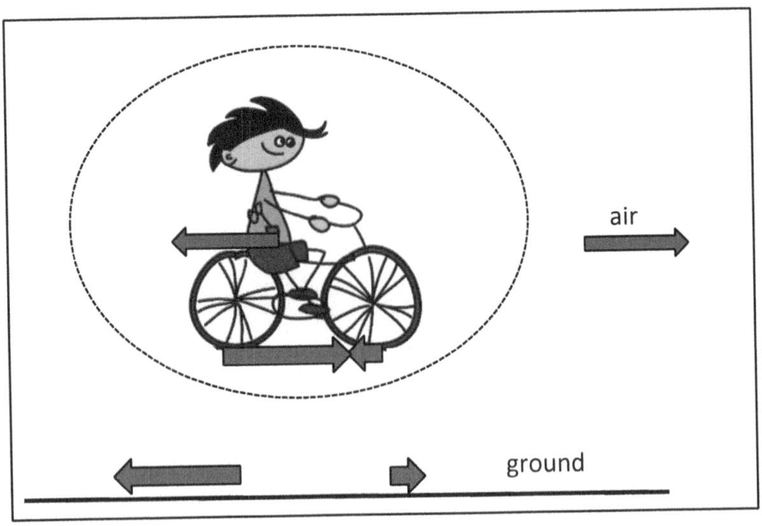

Fig. 2.6 Dislocated diagram to show how the second and third Newton's laws can account for the cyclist's motion. The elements of a balanced free body diagram are shown inside the dotted oval

Respecting Newton's two laws imposes to modify in depth the analysis suggested by the text: "Putting together laws and phenomena" proves to be more difficult, but also more profitable, than what was proposed by the exercise analysed. But even before a complete understanding, it is important to realise that the proposed explanation does not respect Newton's laws and must therefore be replaced by another one.

2.4 The Logical Incompleteness of an Explanation

One can also come to criticize an explanation which does not contain an obvious internal or external contradiction. This is the case for explanations in which at least one link that is necessary for a satisfactory explanation is missing. Of course, no analysis is entirely complete, there are always missing aspects. As Ogborn (1996) writes, "Explanations are like the tip of an iceberg, with a large amount of supporting knowledge lurking below the surface" (p. 65). So when we speak of incomplete explanations, we are not referring to the need to reconstruct all the concepts at stake, those "below the surface". Rather we are focusing on the idea of an essential link that is missing in an argumentative chain and on the case where, logically, something prevents us from reaching a valid conclusion.

The simplest case is perhaps a tautology. Thus, in a popular text (Jacquier and Vannimenus 2005) we find this comment: "In a transparent medium, like glass for example, light propagates more slowly because its refractive index is greater than that of air" (p. 16). In this extract, the expression "refractive index" is presented as an explanatory element whereas this quantity (n) is, by definition, proportional to the inverse of the speed of light in the transparent medium (v) ($n = c/v$ with c: speed of light in the empty space). Presented in this way, the phenomenon to be explained and its "cause" are one and the same thing. The explanation could not be more incomplete.

Less obviously incomplete are the common presentations of tests in chemistry, for example:

> When you put anhydrous copper sulphate on a piece of apple, it turns blue. We know that anhydrous copper sulphate turns blue in the presence of water. Therefore, the apple contains water.

or

> When the gas in the sparkling water comes into contact with clear water, it becomes cloudy. We know that limewater will become cloudy in the presence of carbon dioxide. Therefore, the gas in the sparkling water is carbon dioxide.

In both cases, nothing is said about the fact that only the detected substance would produce the observed effect. Is it only with water that anhydrous copper sulphate turns blue? Is carbon dioxide the only gas with which limewater becomes cloudy? In the absence of answers to these questions, there are clearly missing links in these arguments.

Radiocarbon dating is another relevant, though less simple, example. Most carbon is ^{12}C but a small proportion of it is ^{14}C which is radioactive. Thanks to some documents, found for example on the Internet, we can learn that the technique of radiocarbon dating is based on the law of radioactive decay, which gives the number $N(t)$ of ^{14}C atoms of an initial population N_0 that have not yet decayed after a time t: $N(t) = N_0 \exp(-\lambda t)$. The constant λ is simply linked to time τ ("half life", in this case 5730 years) after which only half of the initial atoms remain ($\lambda = \ln 2/\tau$). It is accepted that at the time of death of a living organism, its carbon skeleton has the same $^{14}C / ^{12}C$ proportion as the atmosphere, due to exchanges related to its metabolism. These exchanges stop at the time of death and then the $^{14}C / ^{12}C$ proportion decreases according to the law that has just been recalled. So far, so good, but how can we know the value of this proportion at the time of death? This information is necessary for dating, i.e. to deduce from a contemporary measurement the time elapsed since the death. The answer is that it is assumed that the composition of the atmosphere has been stable since then, and was therefore the same at the time of death and at the time of dating. At this stage, a demanding mind should ask itself the following questions: why is it that the population of ^{14}C atoms decreases in the dead organism and not in the atmosphere? Is there no radioactive decay in the atmosphere? Something is missing in the explanation (a complete explanation is detailed in Appendix C). In terms of developing a critical attitude, helping students to identify such a defect in the reasoning is a relevant objective. A recent study of 10 teachers at the end of their university preparation shows that it is not self-evident to identify what is missing in this explanation (Décamp and Viennot 2015). These teachers were presented with six explanations of radiocarbon dating of increasing completeness. For each document, the interviewee was invited to state to what extent they were satisfied, or whether they would need further information. The findings indicated that most students (8/10) needed a notable delay before asking the questions raised above, although they had the necessary knowledge to do so.

2.5 What Is Shown by Experiments: Over-Generalisation?

One might think that the outcome of an experiment is the ideal stimulus for a critical analysis. This is one of the basic ideas of "Inquiry Based Science Education", a movement that has generated considerable effort of development worldwide, particularly in primary and secondary education (see for instance Edelson et al. 1999; Flick and Lederman 2006; EU 2007; Gormally et al. 2009; Calmettes 2012). But let us be careful. An experiment that is not consistent with a hypothesis is generally recognized as a serious warning signal, but "confirmation by experiment" must be carefully considered. Generally, an isolated experiment is not sufficient to justify a conclusion. As Bächtold (2012) recently put it, "Experiment underdetermines theory" (p. 21). Bias, experimental error or uncertainty may affect a measured quantity. Researchers make considerable efforts to avoid or reduce these, and we should never lose sight of this concern. But even when there is no doubt at the experimental level, certainty is not yet accessible (this issue is discussed further in Appendix A).

For example, consider an argument like: Steel objects do not float because they are denser than water. Immersing a steel block in water is a striking confirmation of this statement. The steel block sinks; yet it is enough to consider a steel ship's hull to invalidate this argument. An all-steel boat can float if it is properly shaped and launched in the right position. This illustrates the problem of variables that are relevant but not taken into account, such as, here, the shape of the object and how it is put in the water.

With this example, we are not far from the case of energy transfer by aluminium described above. The rapid melting of the ice cube on the aluminium lid of a glass filled with hot water - an indisputable observation - seems to confirm the validity of the statement: "With aluminium we cannot protect against cold", because this metal is a very good conductor of "heat". Here, the radiative properties of aluminium are ignored, whereas, in fact, these properties contribute to explaining the effectiveness of the survival blanket. Ignoring relevant variables or phenomena and over-generalising probably constitute the greatest risks regarding "experimental confirmation", and more widely, scientific reasoning. We will encounter this risk on several occasions. We speak of "functional reduction" when a quantity is wrongly seen as depending on fewer variables than is necessary.

Although it is not immediately obvious, this idea of functional reduction can be used to analyse the confusion between correlation and cause. Believing that an observed correlation between two variables indicates a causal dependence (in one direction or the other) is a line of reasoning that is both frequent and invalid (Gauvrit 2007; Ellenberg 2015). What, for example, can be deduced from a positive correlation between a shoe size above the national average and success in a prestigious entrance exam to a major scientific school? Should young people try to lengthen their feet to increase their chances of passing this exam? Of course, a third variable arises to solve the apparent paradox: gender, which is on the one hand mostly male for winners of this type of competition, and on the other hand, linked to shoe sizes larger than the national average (including women). If one observable was likely to be linked to only one other observable, a positive correlation of their values in a given population could be interpreted in terms of dependence (although we would still not know which is the cause and which is the effect). Otherwise, there is often at least a third variable ("confounding variable"), which influences the first two in parallel.

Even so, it is often difficult to exhaust the critical diagnosis of the variables likely to intervene in an observed correlation: should gender be considered as a major determinant of success in this examination or should sociological factors be included in the assumed determinism? In any case, ignoring the possible existence of confounding variables is a form of functional reduction and may lead to accepting false conclusions.

2.6 When a Line of Reasoning and the Result of a Thought Experiment Are Incompatible

Another strategy that can be used to identify the need for critical analysis is thought experiments, which often involve reasoning about "extreme cases". In a thought experiment, it is not necessary to observe what really happens in the material world to reach a conclusion. For example, if someone says "in a greenhouse, more energy is entering than leaving", or "In a gas, collisions between molecules produce heat", the question of what would happen if this continued indefinitely does not need to be tested experimentally. We *know* that the amount of energy inside the greenhouse would constantly increase, and would lead to an explosion. We also know that if collisions between molecules introduced endless energy into a thermally isolated room (i.e. blocking the transfer of energy to the outside), people inside would not survive for long.

One thing that is not self-evident is about the value of asking this type of question, i.e., with regard to the above examples, of considering the results over time of the phenomena mentioned. This is not common. Instead common reasoning tends to focus on transformations and the causes of these transformations, implicitly considering them to be of limited duration. Permanence is often overlooked (Viennot 2001). But when this effort of imagination is accepted, the need to reconsider a statement like "In a gas, collisions between molecules produce heat" becomes imperative.

To return to a more familiar situation, it is common for second home owners, when they close the house before winter, to wonder if it is appropriate to set the thermostat to the lowest possible temperature (at least a few degrees Celsius, to avoid the freezing of the water in the pipes). A frequent comment is: "It will cost more energy to increase the temperature (from such a low value) when we return". In fact, it is necessary to reason about continuous heat transfers from the "hot" source (inside the house) to the outside. At any time during the owner's absence, these transfers will be smaller if the thermostat is set as low as possible (Fig. 2.7).

But an argument that has more impact is to consider a very long absence in a very long winter. Then, the interest of setting the thermostat as low as possible appears to be a priority with respect to the final warm-up period. This reasoning is considered convincing, but it should not be forgotten that it is worthwhile to set the thermostat as low as possible even when we consider the transients (at the beginning and end of the owner's absence): at any time after the owners' departure, the energy transferred to the outside is less (or equal) with a lower thermostat setting than with the higher one, as shown in Fig. 2.7. The line of reasoning based on an extreme case at least allows critical judgement to be activated.

In fact, the case of an infinite duration is classically used in teaching. In cases where certain functions, or solutions of differential equations, take improbable values for infinite values of time, these functions are eliminated as candidate solutions of the problem to be solved. For instance, exponentials involving time with a positive exponent (e^{kt} with $k > 0$, for instance for the amplitude of a light wave) are said to "diverge" which entails their discredit as possible solutions for the problem at hand. Here too, no experience is required for elimination.

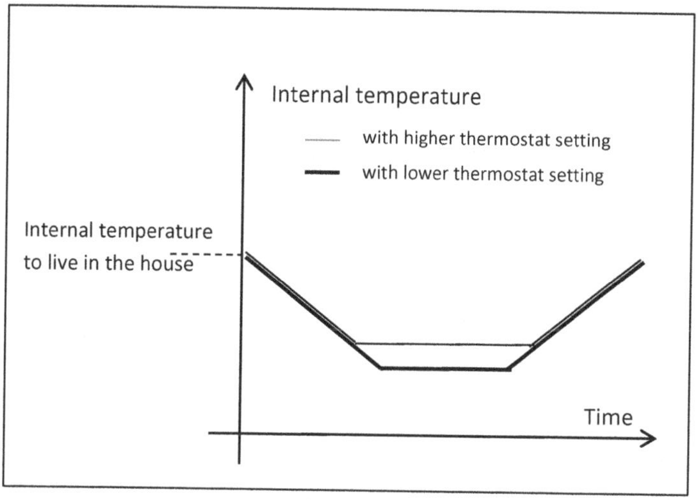

Fig. 2.7 Internal temperature of a house with a higher (fine line) or lower (thick line) thermostat setting: after a given period of absence, the internal temperature is never higher in the second case (lower thermostat setting) than in the first (higher thermostat setting). Consequently, at any time, the energy transferred to the outside during the owners' absence is less (or equal) with the lower thermostat setting

2.7 When an Experiment Should Have Been Performed

However, in the case of thought experiments, i.e. lines of reasoning that do not require real experience, it is wise to be cautious. It may be a mistake to assume that an experimental check is unnecessary. A typical case is the recurrent and erroneous prediction of how the relative ranges of water jets coming out of holes in a bottle depend on the depth of the exit hole (a physical analysis of this phenomenon is offered in Appendix E). The examples in Fig. 2.8 show that the authors of these diagrams had no doubt: the range of the jet on the surface on which the bottle sits *had* to increase with the depth of the exit hole. So they felt no need to check this experimentally.

This is a textbook error that shows a remarkable obstinacy. There are a number of reasons for this. First, a significant trap is believing that a line of argument is valid if the conclusion is in accordance with a well-known fact: here, in a container filled with a liquid at rest, the pressure increases with depth. This statement is correct. There is a second reason for error: the tendency to ignore a second relevant variable, i.e. the free fall time between the exit hole and the surface supporting the bottle. One might think that, with a higher horizontal exit speed, the water will go further. However, to have a large range, the water needs time before reaching the support. A way of seeing that the textbook diagrams must be incorrect is a thinking experiment using an extreme case: if you drill a hole at the level of the support, then it is unlikely that the water will move far from the bottle.

Some textbook illustrations show the jet from the lowest hole having the largest range. According to these authors, this shows that pressure increases with depth.	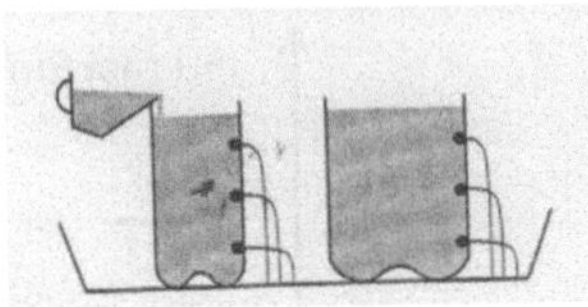 Hibon 1996 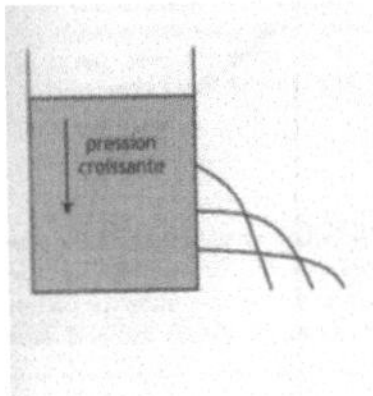 Santamaria 2007 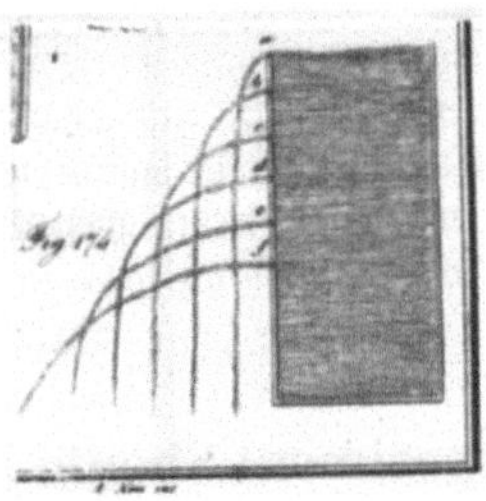 da Vinci 1828
In fact, it is true that the exit velocity increases with the distance from the hole to the free surface of water, but the time of fall increases with the distance of this hole to the surface on which the bottle stands. It is the product of these two distances that determines the range of the jets on the stand. If the values of these distances are exchanged (symmetrical holes with respect to the middle of the height), the ranges are equal.	 Courtesy G. Planinsic

Fig. 2.8 The range of the jets emerging from a bottle with holes

2.8 A Multiple Diagnosis for Useful Calls into Question

The defects we have illustrated above can arise in parallel to raise questions about a given text. It can then be difficult to decide which is the main objection to a text: internal contradiction, (direct or indirect) contradiction of a law, incompleteness of argument, or something else?

This is the case for the following example. Given a closed series circuit with a battery, it is often said or suggested (see Appendix F) that the movement of electrons in the wires is explained by the charges on the ends of the battery (or the field they create), for instance:

> Between the two terminals of a battery there is a permanent difference in the density of the free electrons: the negative terminal has a higher than normal electron concentration while the positive terminal is deficient in electrons. If an electrical circuit is connected to the battery, the free electrons of the circuit are attracted by the positive terminal, repelled by the negative terminal of the battery. They circulate from the negative terminal to the positive terminal outside the generator. (Academy of Bordeaux)

But the field of a dipole follows bean-shaped lines, which is generally not the case of the circuit wires.

If we assume that the local average velocity of the electrons in the wires is parallel to the field, it seems that there is a contradiction. One can even think - a thought experiment - of bending a wire to make electrons flow in a direction opposite to the dipole field (as shown in Fig. 2.9). The question then arises as to why the flow of electrons follows the line of the wire, no matter how it is twisted or bent.

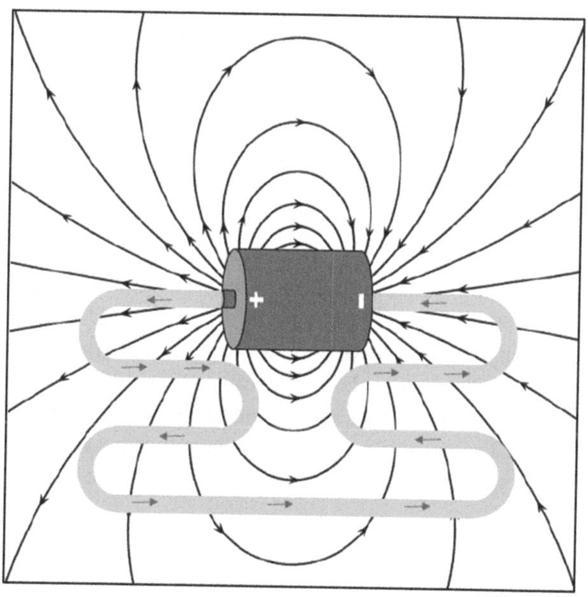

Fig. 2.9 The electric field created by the charges at the battery terminals cannot, on its own, account for the movement of the electrons in the wires

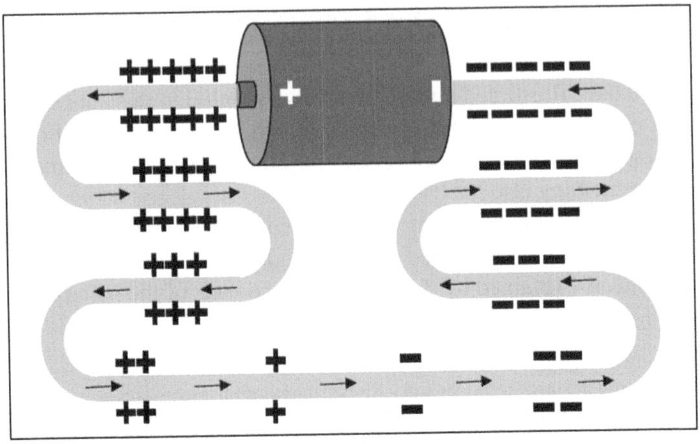

Fig. 2.10 The field in the wires of an electric circuit in quasi-stationary regime: the model of surface charges. Not represented, the charges in the vicinity of the circuit curves are essential to explain that electrons everywhere have an average velocity parallel to the wires

But a defender of the text could also argue that a dipole field is present, even if there are other charges on the surface of the wires throughout the circuit as soon as the circuit is closed (Fig. 2.10, see in particular Chabay and Sherwood 2006).

This lenient defender might conclude that the text is essentially incomplete. That said, what is important is not to characterize the main flaw in an explanation, but to find at least one entry in the critical analysis. It should also be noted that if we consider the movement of the ions present inside the battery, it is in complete contradiction with an analysis based on "polarity" of the battery terminals: in a battery, the motion of cations, which are positively charged, is directed towards the positive terminal: see Appendix F.

That being said, the activation of a critical analysis does not provide the key to the enigma that can then be raised: How is it that the composition of the atmosphere in radiocarbon or stable carbon can be constant over time (see Appendix C)? How can aluminium serve as protection against the cold while being a very good conductor for thermal transfer (Appendix H)? How do electrons manage to take turns in the connecting wires and how cations manage to move inside the batteries (Appendix F)? And so on. It can therefore be considered difficult to tackle such issues without finding a satisfying answer. For the time being, we postulate that, even without having a solution to the questions posed, it is a useful intellectual step to put markers of refusal or doubt on texts whose aim is, in principle, to clarify the understanding of physical phenomena. The point is not to blame the author but to locate the points on which our own reflection must progress: a roadmap for the next stage.

References

Bächtold, M. (2012). Les fondements des sciences basés sur l'investigation. *Tréma, 38*, 7–39.

Belin. (2012). *Sciences physiques* Classe de Terminale S. Paris: Belin.

Brasquet, M. (1999). Actions, interactions et schématisation. *Bulletin de l'Union des Physiciens, 816*, 1220–1236.

Calmettes, B. (dir.). (2012). *Didactique des sciences et démarches d'investigation*. Paris: L'Harmattan.

Chabay, R. W., & Sherwood, B. A. (2006). Restructuring the introductory electricity and magnetism course. *American Journal of Physics, 74*, 329–336.

da Vinci. *Del moto e misura dell' acqua di da Vinci. A spese di Francesco Cardinali*, Bologna (1828). Digitized copy of Harvard College Library, Google books. http://www.archive.org/stream/raccoltadautorii10card#page/n537/mode/1up

Das, S., Marchand, A., Andreotti, B., & Snoeijer, J. H. (2011). Elastic deformation due to tangential capillary forces. *Physics of Fluids, 23*, 072006.

Décamp, N., & Viennot, L. (2015). Co-development of conceptual understanding and critical attitude. Analysing texts on radio-carbon dating. *International Journal of Science Education, 37*(12), 2038–2063.

Edelson, D. C., Gordin, D. N., & Pea, R. D. (1999). Addressing the challenges of inquiry-based learning through technology and curriculum design. *Journal of the Learning Sciences, 8*(3–4), 391–450.

Ellenberg, J. (2015). *How not to be wrong*. London: Penguin.

EU. (2007). *Science education now: A renewed pedagogy for the future of Europe* (European Commission). Brussels: EC.

Flick, L. B., & Lederman, N. G. (Eds.). (2006). *Scientific inquiry and nature of science: Implications for teaching, learning, and teacher education* (Science & Technology Education Library). Dordrecht: Kluwer.

Gauvrit, N. (2007). *Statistiques méfiez-vous!* Paris: Ellipse.

Gormally, C., Brickman, P., Hallar, B., & Armstrong, N. (2009). Effects of inquiry-based learning on students' science literacy skills and confidence. *International Journal for the Scholarship of Teaching and Learning, 3*(2), 16.

Hibon, M. (1996). *La physique est un jeu d'enfant: activités d'éveil scientifique*. Paris: A. Colin.

Jacquier, B., & Vannimenus, J. (2005). *La lumière et la matière*. Les Ulis: EDP Sciences.

Marchand, A., Weijs, J. H., Snoeijer, J. H., & Andreotti, B. (2011). Why is surface tension parallel to the interface. *American Journal of Physics*, 999–1008.

Menigaux, J. (1986). La schématisation des interactions en classe de troisième. *Bulletin de l'Union des Physiciens, 683*, 761–778.

Ogborn, J. (1996). *Explaining science in the classroom* (p. 65). Buckingham: Open University Press.

Santamaria, C. (2007). *La physique tout simplement*. Paris: Ellipses.

Viennot, L. (1982). L'action et la réaction sont-elles bien égales et opposées? *Bulletin de l'Union des Physiciens, 640*, 479–485.

Viennot, L. (2001). *Reasoning in physics the part of common sense*. Dordrecht: Springer.

Viennot, L. (2013). Les promesses de l'Enseignement Intégré de Science et Technologie (EIST): de la fausse monnaie? *Spirale, 52*, 51–68.

Viennot, L., & Décamp, L. (2016). Co-development of conceptual understanding and critical attitude: toward a systemic analysis of the survival blanket. *European Journal of Physics*, 015702 (26pp). https://doi.org/10.1088/0143-0807/37/1/015702

Chapter 3
Risk Factors

The characterization of flaws proposed above can help us to question a text on the basis of various elements of incoherence. But our critical vigilance can be activated even before anything inconsistent or clearly insufficient on a logical level is detected in an explanation. Let us call this the recognition of risk situations. In short, the texts of interest in this book are of two types. Some explanatory element or elements may be directly criticized because they convey unambiguously inaccurate explanations or logically unsatisfactory arguments, such as the examples in Chap. 2; or they may simply suggest misleading interpretations. It is not always easy to distinguish between these two cases, but it is not very important as our point is not to rate the author. The important thing is to be vigilant, not to be misled by an explanation or, if we use it with others, not to mislead them.

3.1 Two Aggravating Circumstances

3.1.1 The "Exactness" of the Conclusion

In a somewhat provocative way, we first designate as an aggravating circumstance for our critical vigilance the fact that a line of reasoning leads to a conclusion considered accurate. This is a very general phenomenon reported by cognitive psychologists: a line of reasoning that leads to a conclusion that is – rightly or wrongly – considered accurate is often, for this reason, considered valid, which is not necessarily true. Kahneman (2012) comments: "This experiment (…) suggests that when people believe a conclusion is true, they are also very likely to believe an argument that appears to support it, even when these arguments are unsound" (p. 45). This is "confirmation bias", often observed with regard to beliefs without any rational basis. But there is no reason to limit this phenomenon to irrational

L. Viennot, N. Décamp, *Developing Critical Thinking in Physics*, Contributions from Science Education Research 7,
https://doi.org/10.1007/978-3-030-43773-2_3

beliefs. It is also observed when an argument leads to a statement that is well accepted by physics experts.

This is the case of an experiment carried out by Lavoisier. Unlike Lavoisier's original text (1789), this experiment has been currently presented (in textbooks at Grade 8 level in France: Boizier 2012) as an exemplary discovery process. The proposed reasoning is based on the observation of the oxidation products of mercury heated in the presence of air (Fig. 3.1). The facts that small red areas appear on the surface, and that the remaining gas (oxygen was used for the oxidation of mercury) does not allow the survival of the mice that are placed there suggest this gas is nitrogen. This conclusion seems entirely acceptable because in fact, apart from a few inert gases, air is a mixture of oxygen and nitrogen in proportions approximately compatible with this experiment. But how do we know that there is no other gas under the bell jar (Fig. 3.1), such as carbon dioxide or mercury vapour? This school use of Lavoisier's experiment could therefore be placed under our heading: logical incompleteness of an argument. In the absence of any other information, this experiment does not allow the composition of the air to be deduced. But since it seems that we are getting the correct result – as we know now – why worry about the possible invalidity of a line of reasoning leading to this result?

Another very emblematic example concerns the condition for the take off of a hot air balloon (Fig. 3.2). If we assume a uniform pressure (i.e. "atmospheric pressure" p_0) inside and outside the balloon, we can easily find the solution that is expected for such classical exercises: For a hot air balloon of total mass m_c (solid parts), given the expression of density ρ in a perfect gas of temperature T and (mean) molar mass M, ($\rho = \frac{Mp}{RT}$, R being the ideal gas constant) and Archimedes' theorem, the Newtonian balance at equilibrium is written: $m_c + \frac{M}{R}\frac{p_{\text{int}}}{T_{\text{int}}}V = \frac{M}{R}\frac{p_{ext}}{T_{ext}}V$ or else,

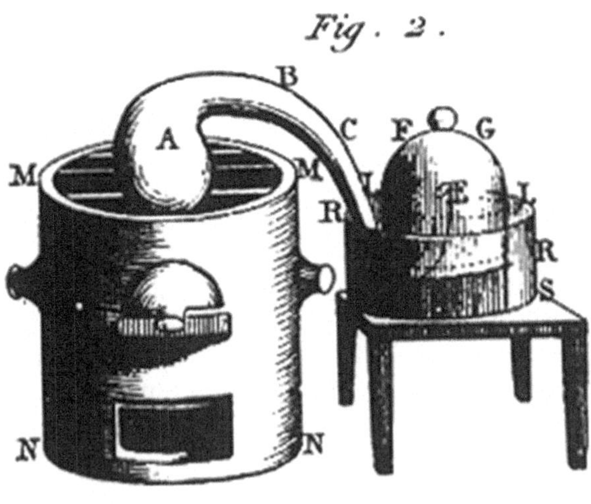

Fig. 3.1 The device used by Lavoisier (1789, part I, chap. 3) to investigate the composition of air

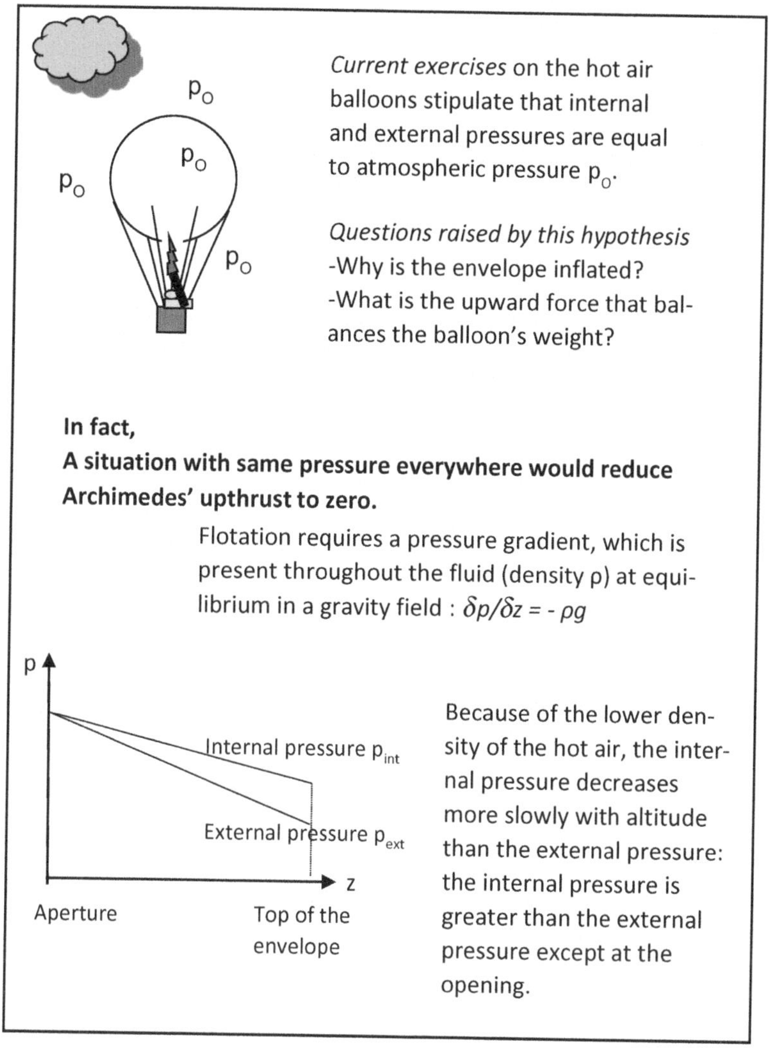

Fig. 3.2 A teaching ritual and its critical analysis (Viennot 2014)

admitting that internal and external pressure are equal to their value at the aperture $p_0 : [1/T_{ext} - 1/T_{int}] = m_c R/(p_0 MV)$.

In fact, the hypothesis of a uniform pressure inside and outside the balloon is absurd because all lift is intrinsically linked to the existence of a pressure gradient. One can add that an upward force on the envelope cannot exist without a pressure difference on its two sides. Moreover, how can we explain why the envelope is inflated?

But using Archimedes' theorem dissolves – so to speak – these objections since, with the above hypothesis, we arrive at the desired conclusion. This does not help us to question the starting point of the argument. Note that we also arrive at this conclusion, which is accurate at first order, when explicitly using the pressure gradients.

3.1.2 Echo-Explanations

It is also commonly observed that explanations that would deserve critical analysis share some characteristics typical of common reasoning. The expression *echo-explanation* is used here to designate this type of resonance which, consciously or not, is activated by the person who produces an explanation. This recalls Jacobi's (2005) remark about public understanding of science:

> The relationships (of authors in public understanding of science) with non-scientific representations in science are more ambiguous. On the one hand, authors intend to change what they think are inappropriate opinions of their readers. On the other hand, they do not refrain from relying on these familiar representations or from transmitting traditional conceptions. In the same way, the rhetorical forms of written discourse are themselves heirs to the tradition of narrative (p. 148).

In what follows, we will encounter several times this proximity between questionable explanations from the point of view of accepted physics on the one hand, and common reasoning in the same field on the other hand. It is not a risky assumption to see this as a reason for mutual reinforcement. It should be noted, however, that what are often referred to as "pre-scientific conceptions" do not always intervene clearly and directly in the questionable arguments we are discussing. For instance, contending that a hot air balloon is isobaric is not the manifest outcome of an initial conception. Other reasons may explain this teaching ritual, firstly the fact that this simplification leads to a quite acceptable value of the internal temperature.

3.2 Simplifications

One obvious risk factor is simplified explanation or excessive use of shortcuts (see for instance Michaut 2014). Oversimplifying an explanation may damage its very consistency. It is, however, worthwhile to elaborate a little on the possible types of simplification and to provide specific examples to illustrate this overall judgment.

3.2.1 Designation of Entities: Not Always Conform to Accepted Physics

When faced with an explanatory or predictive text, the first critical question could be: "What entities are we talking about?" But the concepts involved in a text are not always labelled in the most academic style. Several types of violations are observed. In particular, concepts may take on the status of quasi-material entities, as in this statement found in a book for college students: "From a given drop, only one wavelength reaches the eye of the observer". Of course, in a scientific environment, readers can be expected to understand "wavelength" as "radiation of a given wavelength". For them, the shortcut is harmless.

Sentences that draw upon several notions without clearly specifying their links are also frequently used to define or designate concepts without bothering too much with rigour. Thus this phrase found in a recent popularization book: "The acceleration of the orbital period of the system" (a binary pulsar in this case). There, most people understand that the orbital period decreases and therefore the frequency increases. But in other cases, it is less certain, as with the phrase also extracted from a popularisation book: "A pressure is a force distributed over an area". The main protagonists are mentioned – pressure, force, area – but in disorder, and in any case without any respect for the respective "dimensions" of the pressure and force quantities, that is, for the fact that their numerical values are expressed with different units.

Another classic example is that of the identification by certain authors between mass and energy, via the very famous formula $E = mc^2$ (instead of $E^2 = m^2c^4 + p^2c^2$, p: linear momentum). It is easy to imagine the difficulties that this identification can create for Grade 11 students working on nuclear reactions: energy is conserved despite a mass defect, not to mention the photons whose mass is notoriously zero even though they possess non-zero energy (Bächtold 2014).

Metaphors are also frequently used, either as a shortcut or as a recourse when it comes to a complex notion. In all cases, it is useful to pay attention to how entities are named, in order to locate possible sources of misunderstanding.

3.2.2 *All or Nothing*

The will to explain implies the will to categorize. In teaching, children are told that there are solids, liquids or gases, mirrors that reflect light while other objects diffuse it or let it through, they are told that there are conductors and insulators, etc. It is not a question of objecting to these categories which allow a first description of the material world. Yet, it is useful to keep this process under control. We know, at least one day we learned, that things are not so clear-cut. Glass can both reflect and transmit light, a fact that a passenger on a train can observe when passing from a tunnel to an open-air section. Concerning pigments which are said to absorb, or not, such or such a spectral band, it is enough to direct a laser beam on to them to be convinced that they do not absorb it *completely* (Fig. 3.3) (Viennot and de Hosson 2012, 2015). This distinction between total and partial absorption is both imperative, to respect the coherence between explanation and experimental observation, and of a considerable explanatory range.

Indeed, contrary to what is suggested by the expression "subtractive synthesis of colours", the effect of an absorbent object (for instance a filter or a pigment) on the light consists in *multiplying* its intensity (respectively transmitted or diffusively reflected) by a factor lower than 1. A total absorption is when this factor is zero, which is not currently the case for any black pigment. If its value is, for instance, 0.05, i.e. if the absorption factor is 5%, an intense light – say a laser beam – will be quite perceptible after transmission or diffusion, whereas an "all or nothing" line of reasoning would lead to the opposite conclusion.

a A red laser beam impacts a black pigment	*b* A red laser beam impacts a green pigment	*c* A red laser beam impacts a red pigment

Fig. 3.3 A red laser beam ($\lambda = 632$ nanometers) is directed on various pigments: Contrary to what is predicted by *all or nothing* rules (case *a*, *b*: black/green pigments totally absorb a red light; case *c*: a red pigment totally diffuses a red light), the incident light is *more or less* absorbed/diffused according to the colour of the impacted pigment. (**a**) A red laser beam impacts a black pigment. (**b**) A red laser beam impacts a green pigment. (**c**) A red laser beam impacts a red pigment

It would be easy to find examples in the context of everyday life where an *all or nothing* analysis leads to similar contradictions.

3.2.3 "Small" Assimilated to "Zero"

In the same vein, it is useful to emphasize the risks of "neglecting" quantities that are believed to be "small'. First of all, any expert physicist will say that what counts is knowing what a quantity is being considered "small" relative to. This is true in any field. One million euros is a lot for an ordinary citizen, but it is of little importance in the budgets of many countries.

Thus, in the example of a hot-air balloon, one could say that the pressure variations between the top and the bottom of the envelope are very small compared to atmospheric pressure (ratio of about 3.10^{-3}). This is even more true with regard to the differences between these variations considered inside or outside the balloon. It might seem that all these terms could therefore be neglected in relation to atmospheric pressure. But there are two objections.

One is that, however small, these variations are responsible for the balloon's lift. Indeed, the forces exerted on the top of the envelope can be found by multiplying the pressure difference between inside and outside by an area of about one hundred square meters. (In fact, we have to calculate a surface integral and not a mere multiplication, save in the unlikely case of a cylindrical hot air balloon.)

The other is a matter of physical principle. The expert analysis of the reason why a balloon holds in the air is fundamentally based on the existence of pressure gradients. To deny this *de facto* raises a serious problem of coherence. Even if excessive simplification does not affect the result of a calculation based on Archimedes' theorem, it can prove important for the understanding of the phenomenon we want to teach.

The same is true of current claims about the incompressibility of liquids. How can we understand that the pressure is not the same everywhere in a fluid in equilibrium at given temperature if density is the exact same everywhere? (Besson 2004).

Another example of the issue of assimilating the small to the null has already been presented in Chap. 1, an excerpt from a press article on a classroom resource site that is intended to serve as an inspiration to teachers in the use of documents in grade 10 level in France. It refers to a champion of free fall mounted in a helium balloon to an altitude of 40 km from where, "in the absence of atmospheric pressure", he could launch into free fall (without friction of the air).

The "all or nothing" simplification process is still at work in this example. True, the pressure value near an altitude of 40 km is very low compared to atmospheric pressure at sea level (a few thousandths), but a calculation shows that this pressure is sufficient to ensure the lift of a stratospheric balloon. The rest of the text talks about free fall. To this end, the approximation of zero pressure at the jump position is acceptable, but it is impossible to "put together" this hypothesis and the presence of a balloon at this position.

Note also the credit given to Pascal for his "discovery" of the "empty space" with an experiment where gaseous mercury occupies, of course, the compartment which is so commonly declared "empty".

3.2.4 A Cause Has Been Found: The Only Cause?

In a previous example – a text about an energy transfer in the presence of aluminium (Chap. 2) – we have already been confronted with the coexistence of several causes for a given effect. The observed difficulty was linked to the fact that this plurality of causes was ignored.

Another example is a Grade 9 exercise: "Given that the Earth is flatter at the poles, explain why g is smaller in Kourou than at the North pole" (p. 165). Kourou is at latitude 5° North, so, our planet being flattened at the poles, this city is more distant from the center of the Earth than a place located at the North pole. It seems that the student were expected to conclude that the gravitational field g – i.e. the force of gravity on a unit mass in the rotating frame of reference of the Earth – at Kourou was lower than at the North pole. But, in fact, two phenomena influence the value of g: the dependence of the gravitational interaction on the distance from the center of the Earth and the effect of the Earth's rotation, which increases with the distance (d) from the Earth's axis of rotation (the contribution of rotation to the acceleration in the rotating frame of reference is $\Delta g_{\text{rotation}} = -\omega^2 d$, ω being the Earth's angular velocity). The two effects add in similar proportions to explain the difference observed between the equator and the pole (0,5%). Here, the conclusion of the truncated reasoning (based on a single phenomenon: greater distance from the center, therefore lower value of g) and the direction of the observed variation coincide, which helps to deflect criticism. However, this reasoning is invalid because it ignores a relevant phenomenon – the Earth' rotation – that could equally well be declared the cause of the lower gravity at the equator.

The same type of risk is present in another textbook. Readers are asked (p. 122): "How can we explain the absence of atmosphere on certain planets"? Next, a table provides, for five planets, the mass and gravity value at their surface, but nothing on the temperature, radius or composition of the so-called atmosphere (Table 3.1).

To answer the question asked, the variables in Table 3.1 are not sufficient. To evaluate the probability of ejection of a gas molecule, we need to know (in addition to its mass) its starting point (that is the radius of the planet or the value of g at the planet's surface) and the distribution of the speed of the gas molecules, therefore the temperature and the molecular mass. But the lines in this table suggest a direct (and exclusive?) relationship between the existence of an atmosphere and the value of the gravity field g at the surface: for this reason, we could draw an invalid conclusion without being aware of it.

It can also happen that a single variable explanation seems to be simply a shortcut. Recently, in a program on a French radio broadcast, the operation of floating wind turbines raised a question: How can their concrete platform float? The presenter then explains that, inside the concrete, there is a hole that "creates an Archimedes' upthrust". It is true that floating implies an Archimedes upthrust, but the reverse is not true. In fact, an Archimedes' upthrust occurs as soon as an object is immersed in a fluid in the presence of a gravity field (g). For a given value of g and a given fluid, it depends only on the volume occupied by the platform in the fluid, here water (neglecting the upthrust due to air with respect to the upthrust due to water). With a hollow floating platform, the upthrust is smaller than with the same platform sitting on the seabed (Fig. 3.4). The other force acting on the platform is its weight. Flotation implies that these two forces – upthrust and weight – are equal. The cavity, and thus the lightening of the platform, allows this balance. On the contrary, the radio shortcut is structured like a causal chain: *cavity* → *Archimedes' upthrust* → *flotation*. As a list of relevant elements to explain the situation, the shortcut is appropriate. But as an explanation, that's another story.

Table 3.1 Data provided in a document proposed for Grade 9 in France, which also displays, in the first line, an image of the five planets with very similar radii

Planet	Mercury	Earth	Moon	Mars	Jupiter
Mass of the planet	330×10^{21}	5954×10^{21}	73×10^{21}	641×10^{21}	$1{,}899{,}000 \times 10^{21}$
Value of g at the surface of the planet (N/kg)	2,9	9,8	1,6	3,7	23,1
Presence of an atmosphere	No	Yes	No	Yes	Yes (very rich)

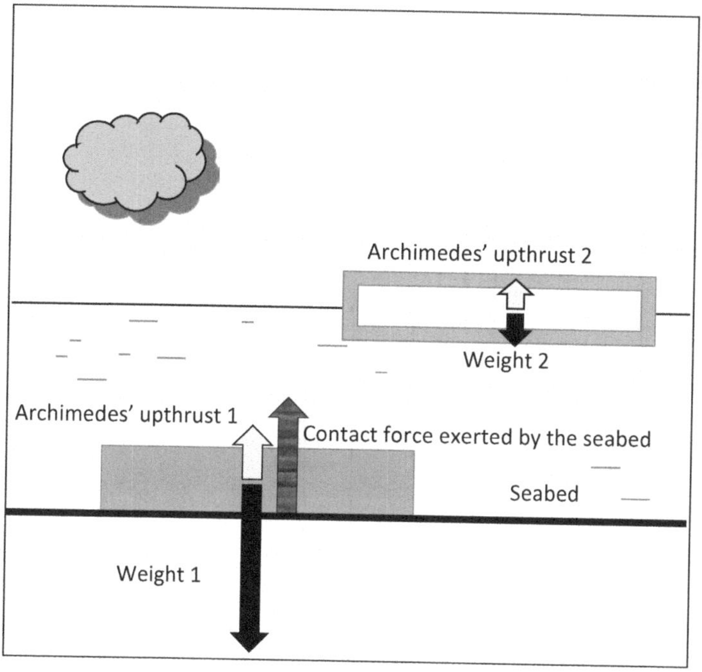

Fig. 3.4 A concrete platform, which floats thanks to the fact that it is hollow (case 2): Archimedes' upthrust is lower than with a platform of same external volume, but not hollow, lying on the seabed (case 1)

3.2.5 A Single Location

In a popularisation text, it is extremely frequent to take into account a single cause for a given phenomenon. This often goes hand in hand with considering only one place, the place where the intended cause-effect relationship appears to occur. For example, Marie Curie commented on the case of a (big!) test-tube, itself containing water, turned upside down over a bowl full of water (Viennot 2014). At the end (the bottom) of the test-tube is the free surface of the bowl at atmospheric pressure. The test-tube, plunged upside down, is filled with water up to 2 m high (Fig. 3.5).

Marie Curie's question-and-answer is: "What is lifting this column of water up by 2 m? It's atmospheric pressure that is pushing on the water in the container. In the tube, there is no air, and no pressure is exerted on the water" (p. 46).

But if this was the case, it would be impossible to reconcile Newton's second law and the idea of a state of equilibrium for this column of water, which is said to be subjected only to its weight (200 N for a section of 0,01m^2, downwards) and to a force exerted by the water at the free surface level (10^3 N for the same tube section, upwards). To ensure the equilibrium, another pressure force is needed, that should be exerted downwards on the liquid column. It is the glass envelope that actually exerts this force (Fig. 3.6). This second interaction occurs at the top of the tube. In

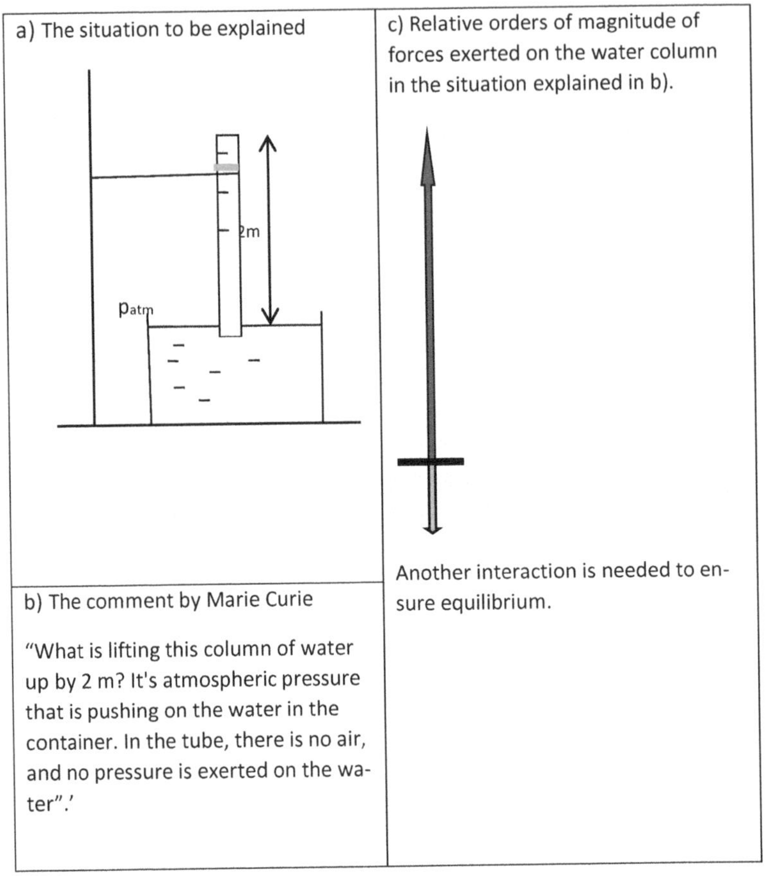

Fig. 3.5 The situation of a test-tube over a container of water, discussed by Marie Curie

Marie Curie's comment, it is ignored as if what mattered was only what served as a support for the water in the column. An inverted glass of water which keeps this water inside thanks to a cardboard situated at its aperture constitutes a very similar situation from the standpoint of physics. Current arguments like the one by Marie Curie also exclusively focus on what occurs under the water column.

Another example of focusing on a single point of an object or a set of objects can be observed in dynamics. It is quite common to think that a force exerted on a point of an object will necessarily accelerate an area close to that point in its own direction. This idea may lead us to believe that the Earth cannot exert a forward force on a pedestrian's foot that accelerates forward, since the contact zone moves (though very little) backwards (McLelland 2011). Except that, for a deformable object, it is quite possible that the center of mass accelerates forward while the contact zone moves backward (Viennot 2014). This is what happens when a runner starts by pushing on a starting-block (which undergoes a very small recoil).

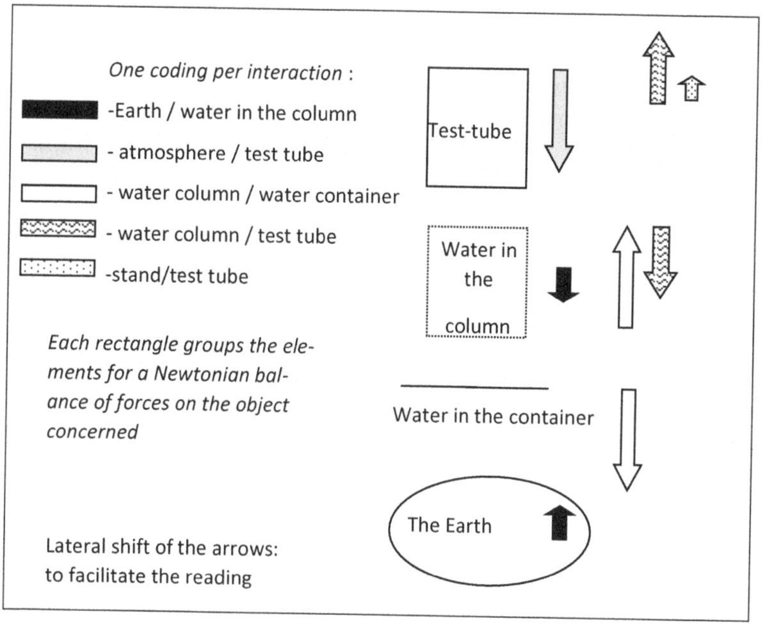

Fig. 3.6 Analysis with dislocated diagrams of the situation of an inverted test-tube (Newtonian balances of forces are complete only for the test-tube and the water in the column)

The elements of simplification mentioned above also intervene in the types of explanation now addressed – *story-like* explanation, *visual* or *analogical* (or *metaphorical*) evocation – and which are sometimes open to criticism.

3.3 The Story-Like Explanation, or "Linear Causal"

3.3.1 Explicit Story-Like Explanation

Let's imagine a text and focus on its structure:

> Someone who is willing to explain something to someone else wants to capture the listener's attention, so they will simplify their speech. This will encourage them to talk about one variable at a time, explaining each time the effect of a change in that variable on another. This entails a linear structure of argumentation, linking binary cause-effect links.

This analysis of a very common reasoning trend actually illustrates what a story-like explanation is. The simple arguments are linked in a linear and causal way. The future tense ("they will…, entails") suggests a chronology in what may seem to be only a series of logical implications. The listener will easily fit into this argumentative structure, as when following a story. In the end, they will rarely care whether, in this example, the linear structure of the text has no cause other than the desire to simplify (for example, the appeal of stories), or even if there is no other way to retain the listener's attention than to simplify.

This argumentative structure and the associated risks have been widely investigated in physics education research (Fauconnet 1981; Driver et al. 1985; Rozier and Viennot 1991; Viennot 2001) and in cognitive psychology (Halbwachs 1971; Kahneman 2012). In physics education research, it has been called "Linear causal reasoning". This expression designates a linear chain of implications, each link mentioning a single phenomenon (ϕ), relating to the evolution of a single quantity: $\phi_1 \rightarrow \phi_2 \rightarrow \phi_3 \rightarrow \ldots \phi_n$. This brings us back to a form of reduction in the number of variables, or "functional reduction", already mentioned.

This is in line with a comment by Kahneman (2012): "Effort is required to maintain simultaneously several ideas that require separate actions or that need to be combined according to a rule" (p. 36). In this respect, functional reduction brings a certain facility. A more specific characteristic is the status of the arrow which is used to symbolize the relationship between two successive phenomena in the explanatory sequence. The arrow between the predicates could represent a logical implication, that is, "therefore". Or it could indicate the occurrence of a subsequent event: "next". In fact, using the "then" connector offers a comfortable way of not choosing between a logical implication and a chronological sequence (Table 3.2).

To sum up, in linear causal reasoning.

- the envisaged events are often described simply, using a single variable;
- more or less explicitly, these events are understood to be successive,
- and hence as temporary, or at least they are temporarily considered.

What disappears in this narrative type of explanation is simultaneity and steady behaviour. This type of reasoning is not appropriate for the so-called "quasi-static" analysis of systems, which consists in assuming that the values of the various variables all change at the same time, while continuously satisfying some simple relationships.

Imagine, for example, a system of two springs (of known un-stretched length and stiffness) end to end suspended from the ceiling; the bottom end is gently pulled. The system can be described by several variables (spring length and tension, total length, external force on the lower spring). These variables are connected by simple relationships, for instance those relating the total length to those of each spring

Table 3.2 Different languages, same ambiguity in meaning: the intermediate term evokes both a logical and a chronological sequence

Status ↓	French	English	Spanish	Italian
Logical	donc	therefore	por eso	quindi
Intermediate	alors du coup	then	entonçes	allora
Chronological	ensuite	later next	despues	poi dopo

(Fauconnet 1981): The question to be resolved is that of the displacement of the junction point for a given total extension of 10 cm (Fig. 3.7). A pupil (in grade 10) writes: "The first spring will extend. Then, after a while, the second will also extend" (p. 112). This comment betrays a sequential vision of the events. Indeed, if we imagine that the extension of 10 cm concerns only the lower spring, if we deduce from that, given its stiffness, the external force acting on this spring, and if we then consider this same force as acting on the higher spring, we are induced to

In quasi-static physics	**An example**	**Linear causal stories**
- several variables - change (almost) simultaneously - constrained by permanent relationships	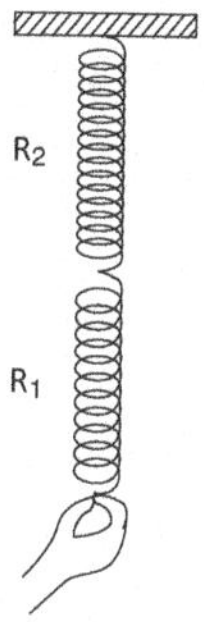	- simple phenomena (one variable each) - are taken into account sequentially (hence) - temporarily
$F_{ext}(t) = T_1$ *(same t)* $= T_2$ *(same t)* $\Delta l_T(t) = \Delta l_1$ *(same t)* $+ \Delta l_2$ *(same t)* F_{ext} : Force exerted by an experimenter on the lower end ; T_1, T_2 : tensions of each spring; Δl_1, Δl_2 : extensions of each spring, Δl_T total extension= 10cm. $k1 = 3N/cm \quad k2 = 2N/cm$ Correct answer : $\Delta l_2 = 6cm$		*A symptomatic comment:* "The first spring will extend then, after a while, the second will also extend." The pupil's calculation $F_{ext} = k1.\Delta l1 = 30N$ $T_2 = T_1 = F_{ext} = 30N$ $\Delta l_2 = \frac{T_2}{k2} = 15cm$

Fig. 3.7 Contrasting a quasi-static analysis and a story-like explanation (linear causal) the example of two springs placed end-to-end

overestimate both this force and the displacement of the intermediate point (we find 15 cm). To find the accurate value, it is necessary to take the *two* springs *permanently* into account. It is because they are both involved that it is not necessary to pull so strongly to reach a total extension of 10 cm.

Of course, one may wonder how it is that, *as soon as* an external action is applied, the higher spring is "warned", so to speak, that its tension must change while respecting the relationship of equality between the two tensions. The validity of this approximation relies on a comparison between two scales of time: for a very slow extension of such a system, the time of propagation of contact interactions along the system is quite negligible with respect to the typical duration of the system extension. In accepted physics, this means that each step of the transformation (then "quasi-static") can be seen as a state of equilibrium, hence the use of laws that are valid for equilibrium.

Thermodynamics provides many examples of quasi-static analyses where several relevant variables are said to change *while constantly* respecting the laws of thermodynamic equilibrium, such as, for an ideal gas, $pV=NkT$ (p: pressure, V: volume, N number of molecules, k: Boltzmann constant, T: absolute temperature) (Rozier 1988; Viennot 2001, 2004). A text (Maury 1989, p. 87) that explains the low pressure (p) at altitude by argumenting that there are "fewer molecules" (by unit volume: N/V) ignores the possible effect of temperature (T). This makes the explanation incompatible with that of a hot air balloon, where there are fewer molecules (by unit volume) inside than outside but where there is no less pressure (at a given altitude) than outside. More generally, the binary causal links that we are tempted to assert regarding an ideal gas are quite inaccurate if nothing is said about the third relevant variable (considering N as constant).

In everyday life, as soon as multiple aspects of a situation permanently interfere, for instance in economics, it is not possible to reason about a possible link between *two variables* while ignoring how the other variables intervene in their interactions with the first two.

3.3.2 A Suggested Story-Like Explanation

Sometimes the explanation is not as explicitly sequential as in the previous examples, rather this aspect is suggested by a text or a diagram.

When we formulate or read a text, we necessarily adopt a sequential format, which may be sufficient to resonate with a linear causal approach to a phenomenon. Even the *and* link readily suggests chronological succession, or a causal link, or both. This is probably what happened when Marie Curie presented a siphon explanation to her young students (Chavannes 1907 2003): "The water contained in the long arm of the siphon flows out. A vacuum is created and atmospheric pressure causes the water to rise in the short arm, which is immersed in the container".

In this text, the vacuum to which reference is made seems to be a consequence of the flow of water out of the open tube on one side and the equally direct cause of the rise of water on the other side. More broadly, we recognize this sequence of simple and localized arguments, starting where the liquid flows (but why does it flow?) and ending on the other side, where atmospheric pressure pushes the liquid up (but the same pressure also pushes the liquid up on the other side, where it flows). The completeness and consistency of the explanation are strongly affected. In fact, we might first imagine that the tube is clogged. Depending on the respective heights of the water level in the vessel and the end of the tube, the pressure in the water at this end is higher (case a) or lower (case b) than the atmospheric pressure. As soon as the tube is opened, the movement of the water/air interface is controlled by the difference in pressure between the two fluids at this point: downwards (case *a*) or upwards (case *b*) (Fig. 3.8).

A colon (:) can also suggest a cause-effect link, for example in this text: "At the summit of Everest (8.848m), the air is rarefied: the attraction of the Earth is 9.760 N/kg instead of 9.811 N/kg at sea level" (p. 122). This implicit causality can lead to ambiguous or clearly erroneous views. In fact, the scarcity of air is neither the cause nor the effect of the lower value of g at altitude. The law of fluid statics stipulates that, in a gravitational field, variations in pressure (p) and altitude (h), for a fluid of density ρ, are related by the relationship $\Delta p = -\rho g \Delta h$, so that any non-zero value of g (whether constant or not with altitude) implies a pressure (and density) decrease with altitude). Even in a constant gravitational field, the air would still be thinner at altitude, and the pressure in water increases with depth in a pool, even in the absence of any significant variation in g.

Compared to texts, images have the advantage of not imposing a sequential reading. They are relatively conducive to the simultaneous evocation of multiple factors likely to influence a phenomenon, even if a reading from left to right is often

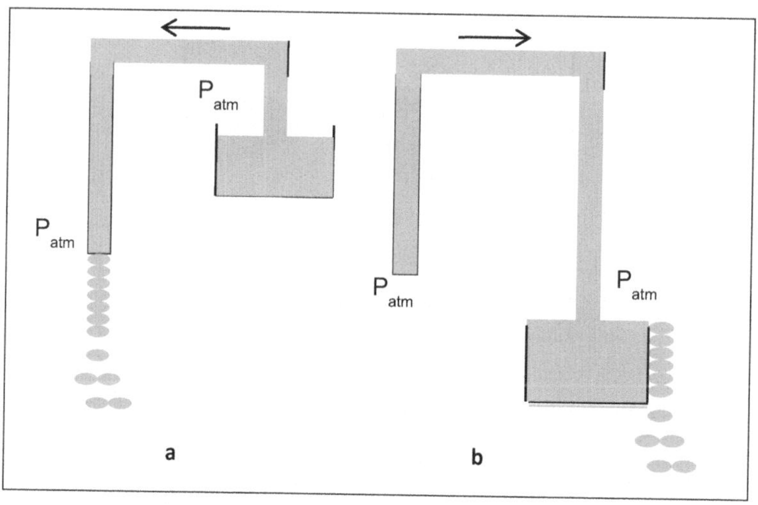

Fig. 3.8 The working of a siphon depends on the respective heights of the free surfaces at the two ends of the system, and *not* on atmospheric pressure

preferred. Therefore, they may protect us from linear causal analysis. But they can also suggest its characteristics, as in the following example on the Hall effect.

If we consider a flat, rectangular and conductive sample traversed by a direct current and located in a perpendicular magnetic field (as in Fig. 3.9a), there is a potential difference between the sides parallel to the magnetic field and to the direction of the charge carriers.

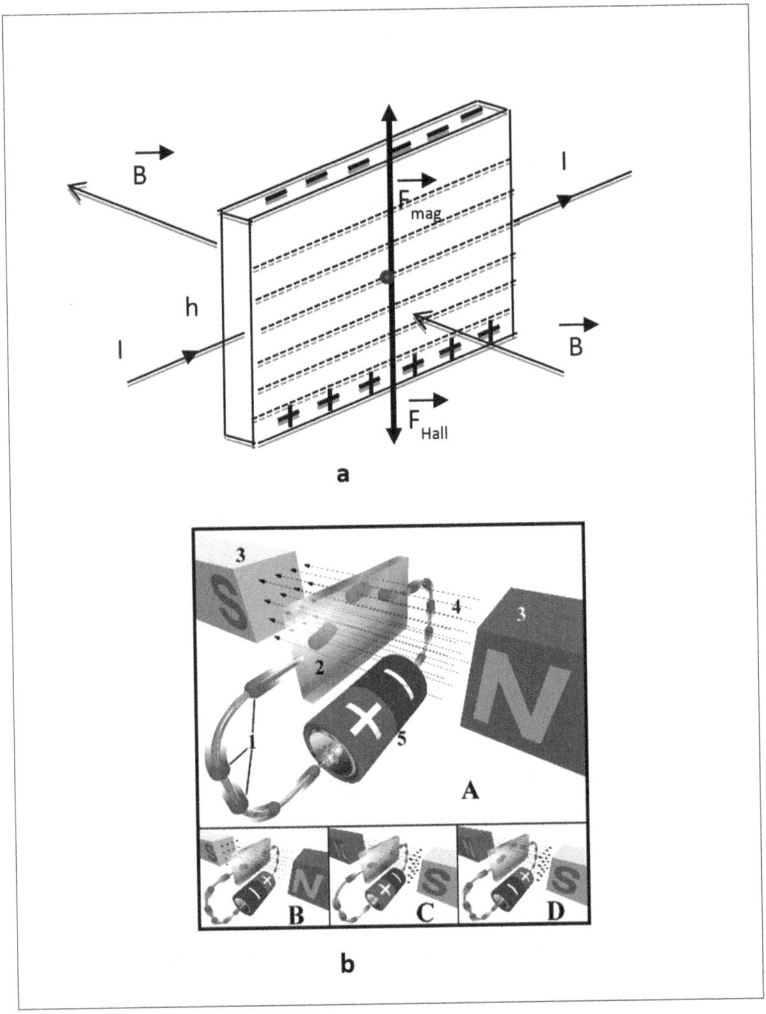

Fig. 3.9 Diagrams showing the movement of charge carriers (here electrons) in a metal when located in an electric field and in a magnetic field ($\vec{F}_{\text{Hall}}$ and $\vec{F}_{\text{mag}}$ are respectively the transversal electric and magnetic forces acting on an electron; (**a**): in steady state, the charge carriers follow the same direction as without a magnetic field; (**b**): an image on Wikipedia which suggests a steady state that would remember a transient: https://fr.wikipedia.org/wiki/Effet_Hall

A few years ago, one could see on Wikipedia a very suggestive diagram showing the path of electrons in a sample: These electrons first follow a direction parallel to that of the current in the absence of a magnetic field, then they curve their path towards one of the lateral faces but do not reach it; rather they return to their initial direction just before leaving the sample (Fig. 3.9b). This trajectory seems to bear direct witness to the history of these electrons. The electrons (charge $q = -e$) first enter, with velocity $\vec{v}$, the area where the magnetic field $\vec{B}$ is present. They are subjected to a magnetic force $\vec{F}_{mag} = q\,\vec{v} \times \vec{B}$ which laterally deviates them towards one side where they accumulate. This creates, between opposite sides, an electric field ($\vec{E}_{Hall}$) which is in the plane of the sample and directed perpendicular to the initial direction of the electrons. This field is associated with a potential difference between the two sides of the sample. The electrons are then affected by an electrostatic force ($\vec{F}_{\mathrm{Hall}} = -\,\mathrm{e}\,\vec{E}_{Hall}$) which sends them back into a median trajectory, then they emerge along the axis.

In fact, this account is not valid for a steady regime. If electrons are deflected to one side, that is, they have a velocity component perpendicular to that side, it means that they accumulate there. But this accumulation is limited by the electrostatic repulsion it causes. The accumulation stabilizes when the magnetic force exactly compensates for the electrostatic repulsion, which happens in a very short time. Then the electrons "move in a straight line" (on average). The history of their sinuous course in Fig. 3.9b is the geometric visualization of a chronological sequence of two phenomena which are, in fact, simultaneous and compensate in the steady state: the electrostatic repulsion due to the accumulation of charges on the sides, and the magnetic force due to the magnetic field.

Permanence (of the situation in steady state) and simultaneity are not easy to envisage. For example, it is very tempting to say that, in this story, everything *begins* with the magnetic force.

The common point to these stories is that several phenomena involving several factors (electric and magnetic forces for Hall effect, volume, pressure and temperature for a gas, first spring and second spring, water pressure on one side and water pressure on the other side for a siphon) intervene simultaneously. These factors contribute to certain oppositions which, in the case of quasi-static regimes, lead to various compromises (electrons in Hall effect "move straight", *both* springs lengthen while keeping– approximately – equal tensions). The value of being warned of this is that, as regards complex phenomena (the whole of physics!), we can be alert when an explanation is patently linear causal.

This is not only true in physics. Economics, for example, involves complex interactions between multiple variables. We are always surprised to hear such and such a speaker say that everything will be better if the national debt is reduced (and consequently…, it implies that…), while another will explain that everything will be better if the purchasing power of the citizen is increased thanks to the reduction of taxes (and therefore…, consequently.., it implies that…). These completely opposite linear causal arguments seem all indisputable and all predict an increased well-being for citizens. Nothing is said of the necessary compromises, of a balance to be found between multiple constraints (monetary value, exports, etc.). So one might think that these arguments could support a shift to the limit: Should taxes be completely eliminated? This is a good question, the kind of question a physicist might ask.

Narrative explanations provide some comfort to the recipient. We have just discussed the associated risks. But given their frequency in popularisation contexts (Viennot 2007), we can conclude that the audiences are pleased with them: everybody likes stories. This is a typical case of echo-explanation, given that the structure of such arguments resonates strongly with common reasoning.

3.3.3 *Story-Like Explanation: When the Risk Is Actual*

One example shows how knowledge of a risk can anticipate the likely effect of an explanation on a given audience, even with text that contains no erroneous statements. Students in higher education (year 1 to 3, $N_T = 181$) were asked to read the following text (Valentin 1983):

> The mean energy of motion of each molecule is enough to prevent the molecules of gas that surround us from being bound to each other: in a gas, molecules are continually colliding and rebounding. But, if the temperature is lowered, the system will be able to become liquid and even solid. These physical phenomena occur when, with decreasing temperature, the kinetic energy of the molecules becomes so low that they can no longer resist the electromagnetic interaction. They first gather in the liquid state and finally get bound in the solid state (p. 13).

Taking into account the preceding remarks, this text could raise a concern: the structure of the explanation, a linear causal story, could lead to an erroneous conclusion concerning the respective values of the average kinetic energy of molecules in the liquid or gaseous phase.

To clarify this point, students were then asked to read the following passage:

> At any given moment of the liquefaction, the mean kinetic energy of a molecule of gas is larger than the mean kinetic energy of a molecule of liquid (liquid and vapour are in thermal equilibrium at the time considered).

Students were asked if they thought the text suggested such a statement and if the statement was true. An overwhelming majority of students considered that the text suggested such a statement (77%, N = 181) and that the statement was true (80%). However, on the next page, the author points out that in thermodynamic equilibrium, the average kinetic energy of molecules is the same in the liquid and gaseous phases. The statement that the students were asked to evaluate is therefore false and was not intended to be conveyed by the text. Yet the students thought that was what they had read. How should this result be interpreted? It may simply be that, because the students thought the statement was true, they thought they had read what they already considered correct. On closer inspection, we see that the passage is punctuated with a temporal connotation: *If… will be able to… when… so low that … they can no longer … first … finally*. This structure can be represented as a logical chain.

$\phi_1 \rightarrow \phi_2 \rightarrow \phi_3 \rightarrow \ldots \phi_n$. The common tendency to insert time into explanations is encouraged by this. The text can be read as a story. First there's a gas, then a liquid. Meanwhile, the temperature, then the average kinetic energy of the molecules, decreases. This quantity is therefore lower at the end of the story than at the beginning. In the statement, the two phases are in equilibrium, i.e. are present simultaneously, at the same temperature, therefore with the same average molecular kinetic energy.

The students' responses suggest that the chronological structure they see in the text overshadowed the idea of simultaneity that the excerpt was intended to convey; this apparently leads to the belief that mean kinetic energy is lower in one phase (the "final phase") than in the other (the "first phase").

In any case, we cannot remain indifferent to the very large rate of erroneous readings, which strikingly echoes the anticipation that inspired this investigation.

That said, a linear causal structure is not the only factor that can cause a distorted reading of an explanatory message.

3.4 Visual Messages or Analogies: The Risks of Showing or Strongly Suggesting

3.4.1 Experimental Evidence

In terms of text interpretation, a risk is linked to the wish to show. Here, we do not address the general question of the validity of an "experimental demonstration" (see Appendix A), but center on the visual aspects of an experiment. "Seeing is understanding" is a popular refrain. But we should be very cautious before accepting it. First, consider the circular presentation of this question proposed by Euler (2004): "You understand what you see" but also "You see what you understand" (p. 193). A look is not passive, it chooses its direction, it disregards many things, according to the intellectual activity or the affective state of its author. It may happen that a person simply does not see what we would like her to see, or says "You have cheated" when presented with the outcome of an experiment (Maury et al. 1977, pp. 15–25). It may also happen that she sees what she is intended to see but without the expected benefit in terms of comprehension. For example, "visualizing" rays of light by dispersing some dust in the path of a laser beam can reinforce the erroneous idea that light is visible from everywhere like a passing train. It is also tempting to use a "raybox" to "show" the rectilinear propagation of light, and to forget to criticize this approach (Fig. 3.10, see also Kaminski 1989). As for our students, happy to *see* (so they think) the rays studied in their lesson, they are even less encouraged to call into question the meaning of the luminous tracks they are observing. Of course, texts reporting some "light visualization" experiences and showing beautiful photos of them share the same risks of misinterpretation.

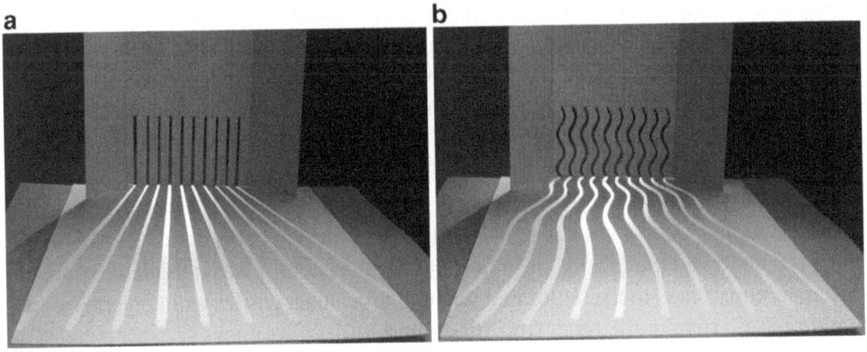

Fig. 3.10 (**a**): A small lamp behind a screen in which parallel slits have been cut produces traces of light on the surface; arrangement (**b**) avoids oversimplification in this respect. In both cases what is seen is a set of shadows: The observed traces are not rays. (Photos W. Kaminski)

3.4.2 Explanations Based on Images or Diagrams

Using images or diagrams as basis of an explanation is no less risky, as many investigations showed (Kress and Van Leeuwen 1996; Gauvrit 2007). Whether or not in line with the author's intention, an iconic representation may inadvertently introduce some elements that may transform, for the receiver, the meaning of the intended message. As with verbal statements, an iconic message can be clearly incompatible with accepted physics, but an image can also, without being strictly speaking "erroneous", induce the reader to perceive a message which does not conform to expert physics. It is important to detect this last case, in particular for a teacher who might use such a document in class. In no way should we refrain from using images or diagrams in teaching, useful as they are to complete our explanations. But it is important to know the related risks.

A few risk factors are outlined below (Pinto et al. 2000; Colin et al. 2002).

3.4.2.1 Realism and Symbolism

The goal of an image is to show, and it may seem strange to call this goal into question. Yet there is a risk that some elements intended as symbolic turn out to be seen as quasi-material, like an ordinary object. Bachelard (1938) pointed out the "substantialist obstacle" which is, in his view, inherent in common thought and consists in seeing concepts as ordinary objects, with a possible explanatory function: "Electricity is a glue" (p. 119). Clearly, a suggestive iconic document may resonate with this trend. Think, for instance, of these rays of light drawn in white on a grey background, as in many textbooks (e.g. Crawford 1965).

When they are drawn as behind a magnifying glass to show their diffusion by a rough surface (Fig. 3.11, Karplus 1969), it may happen (Colin et al. 2002) that

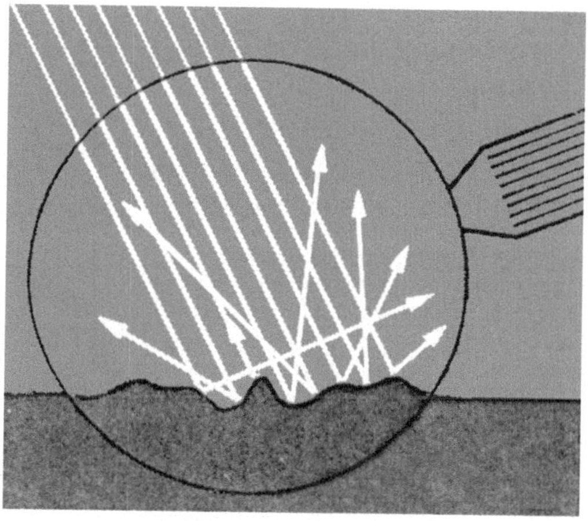

Fig. 3.11 A diagram for the diffuse reflection of light by a rough surface (Karplus 1969, 124)

teenagers conclude that the magnifying glass is there "to show the trajectory of the rays that are extremely small and that we cannot see" (p. 317).

These pupils are far from understanding that a ray is a model used to account for the propagation of light (in a limited domain of situations), but not a kind of luminous spaghetti that could be visible from everywhere.

Sometimes, in popular accounts, concepts are quasi personalized, as is the case for small electrons, more or less tired, that show us the adventures of electric current in a circuit. It also happens that an object is shown to point to an action but without its actual working being taken seriously. For example, a magnifying glass is situated in the foreground of an advertising image and a young observer is represented partly behind it without the least deformation, as if the magnifying glass was there only to help him to see, therefore for only one direction of the light (Fig. 3.12).

3.4.2.2 Similarity of Symbols for Different Entities

Sometimes, when the symbolic toolkit is too limited, a single symbol is used to serve two different meanings. This is typically the case for arrows. Any teacher of mechanics finds it difficult to represent on the same diagram velocities, accelerations and forces without increasing the risk of confusion between these quantities, if not of inappropriate addition of a force and a velocity, which is often observed in students at upper secondary or college level. Using different colours is a must!

Another case of similarity of symbols is shown in Fig. 3.13, concerning, this time, a line of sight and the direction of the light.

It is no wonder that a student in grade 12 concluded: "CD is in the shade but so is AB." (Colin et al. 2002, p. 320), all the more so when we know how readily common reasoning can confound the line of sight and the path of light (de Hosson and Kaminski 2007).

Fig. 3.12 Excerpt from an advertising page for scientific workshops (Reflets de la physique 2017)

Another prototypical example is that of interference with Young's slits (Fig. 3.14).

This type of representation swallows, so to speak, the difference between a ray of light in geometrical optics (useful for analysing what happens before the slits) and a path that is useful for evaluating the phase of the luminous information reaching a given point beyond the slits (Colin and Viennot 2001).

3.4.2.3 Over Selection of Represented Elements

The goal of an image is to show, but, to this end, the designer has to select what will be useful for the intended message. To highlight some elements, it is necessary to disregard some others. Yet that is a possible cause of misunderstanding and it will be necessary to negotiate between such a risk and a complexity that would be counter-productive. Figure 3.14a also raises this objection, because it does not show the "paths of phase" that would be useful for analysing the luminous state at *anothe*r point on the screen than that those shown.

What would be the value of representing more than two of these emerging paths of phase, at least at the start of a teaching session about interference (Fig. 3.14b)? That of avoiding, for instance, students coming to the view that the light is "deviated" by each hole – a notion of geometrical optics – whereas it is a diffraction process which intervenes and permits luminous information to go in multiple

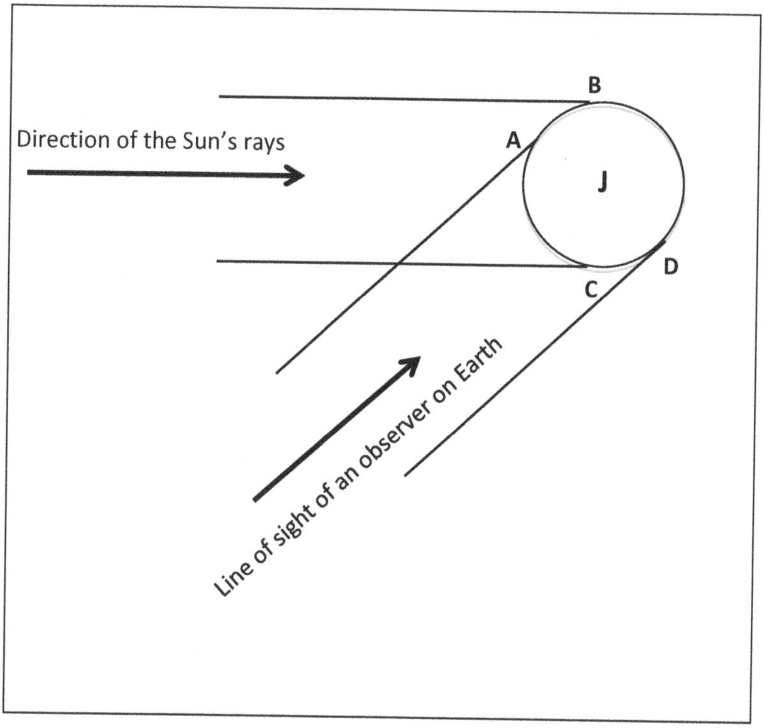

Fig. 3.13 Jupiter seen from Earth: the same symbol is used for a line of sight and the direction of light (Very simplified version of an image shown in: Botinelli et al. 1993, p. 121)

directions: it is only the choice of a given arrival point that legitimates the selection made (This is a "backward selection": see Colin and Viennot 2001).

It is noteworthy that this idea of a "deviated ray" aligns with the narrative of a well identified protagonist – the ray – pursuing its lonely way after an obstacle, knowing where it had to go. This is not far from a story-like explanation, and an image like Fig. 3.14a does not help us to take some distance from it.

A second example relates to colours. It concerns an image that has been used in a previous investigation (Colin et al. 2002) (Fig. 3.15). It is slightly less frequently used in textbooks nowadays.

So we can see this diagram as an example of both over-selectivity of the image – since instrumental contexts are missing – and of similarity of symbols. Circles occupy the two parts of the figure, though nothing prevented the designer from representing filters as rectangles, for example. A comment by a grade 12 student is typical of the ensuing misunderstanding: "Additive synthesis consists in superimposing red, green and blue filters to get white where the colours cross "(Colin et al. 2002, p. 327). A black background would be more consistent with the idea of light addition, since it would show where no light arrives. This image, therefore, constitutes a risky situation for the explanation of additive and subtractive syntheses of colours.

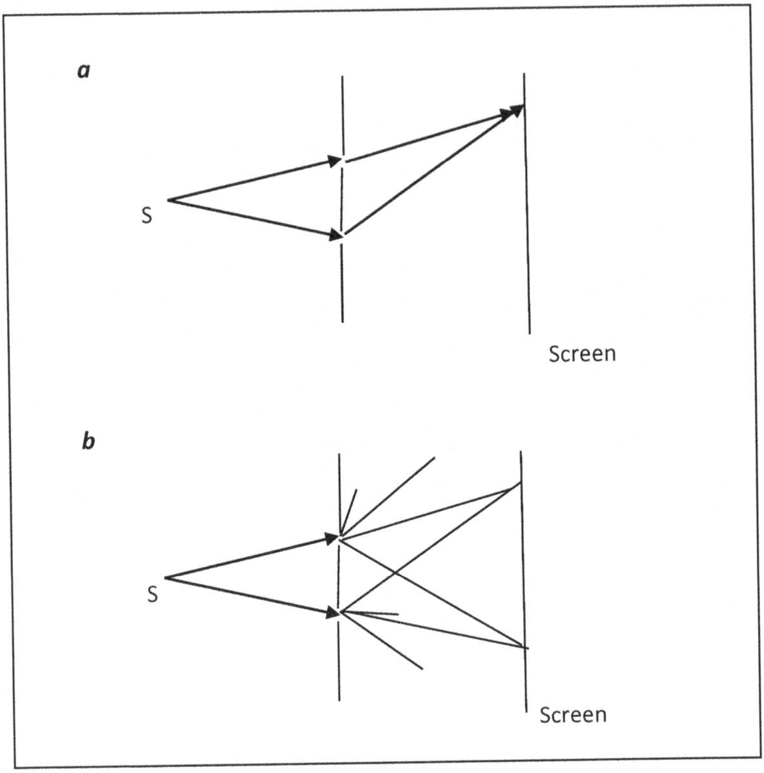

Fig. 3.14 A typical diagram for the Young's slits experiment (**a**) and an alternative representation (**b**) which avoids the similarity of symbols and underlines the backward selection of the paths used to calculate phase differences

An aggravating factor is the all-or-nothing status of these rules, as underlined above (Fig. 3.16). In the absence of any information about spectral intensities of lights or on spectral absorption curves of filters or pigments, we may (erroneously) understand, for instance, that *any* green pigment *completely* absorbs all the red light. The absorbing power of any pigment, including black, is always limited: Illuminated with an intense beam of light, a black object will diffusively reflect enough light for the impacted zone to be visible. Even with an ordinary lighting, we can distinguish the three dimensional shape of black objects, which means that they send us some light. The painter Pierre Soulages is well known for this use of black colour, which he calls '*black-light*'.

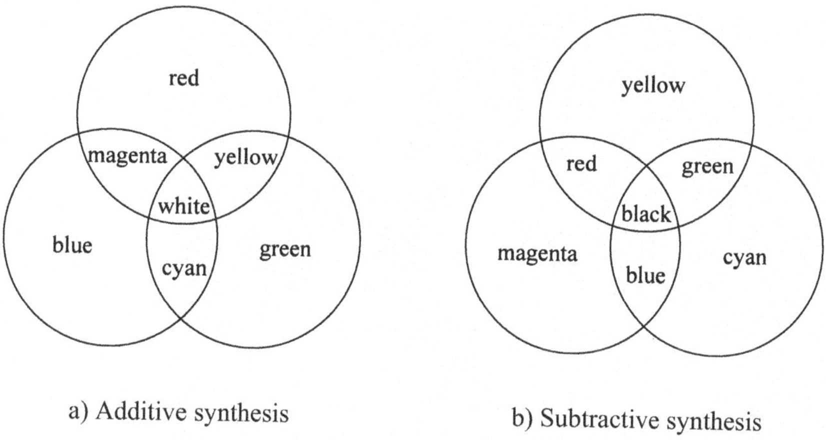

Fig. 3.15 A common representation of additive and subtractive synthesis rules; instrumental contexts are missing (see text)

3.4.2.4 The Structure of the Image and the Scales of Representation

The composition of an image also conveys some information, more or less unintentionally. This is one of the most emphasised aspects (Kress and Van Leeuwen 1996) in research literature about the analysis of iconic messages. In advertising, it is clear that the composition of an image is quite crucial. But it is perhaps less self-evident that images relating to science may convey the same type of implicit messages.

A few years ago, in the *Accompanying documen*t for the physics and chemistry syllabus for grade 10 in France (Centre National de Documentation Pédagogique 2000), in a section on *orders of magnitude*, two extra-terrestrials ('ET') are shown, contemplating the Earth, its continents and its blue oceans, all this on a black background (Fig. 3.16). With their telescope, they can see two hairy humans trying to make a fire. Students are asked to understand how these ETs can deduce their distance from the Earth, given that they have access to this prehistoric scene (about 300,000 years before the ET's discussion) thanks to the propagation of light.

A discerning analyst, however, might wonder how, from their planet, they can distinguish continents and oceans on the Earth, despite their distance of, say, 300,000 light-years (3.10^{18} km). During an experimental study on how to accompany students towards a critical attitude (Feller et al. 2009), the participants (grade 10 students) had collectively reached enough maturity not only to criticise the image but also to consider the standpoint of the person who drew it, wondering how s/he should represent the Earth for us in a recognizable way. This constitutes a valuable step, one which might be taken as an objective in an education in critical analysis at various competence levels.

Fig. 3.16 A suggested document to accompany the grade 10 program in France (Centre National de Documentation Pédagogique 2000)

3.4.3 «Evidencing », Via an Analogy or a Metaphor

At first sight, regarding explanations in physics, the use of an iconic representation is very different from that of an explicit or suggested analogy (a metaphor, in the latter case). Yet these two strategies have some similarities. In both cases, the goal is to facilitate conceptual access to a non-obvious domain of physics via a more directly perceptible evocation, even though, where analogy is concerned, the *evidence* is assumed to stem from a mental representation, without any visual mediation.

Analogy between a situation to be explained and another one that is better known to the listener or the reader is a very popular tool in explanations. This is the case, in particular, when the targeted audience is novice, but not only then. A metaphor, being an implicit analogy, has some similarities as regards the benefits and risks. In both cases, a greater familiarity with a "source" situation is expected, so that we could easily transpose what we know of this "source" situation to the "target" situation, to borrow a frequently used terminology (Gentner 1983). Some explanatory elements are made conceptually accessible, not via an image but via a mental representation which is, in principle, readily available. That being so, it is well known that, even if there are some common features or an identical structure between two situations, drawing a parallel always has some limits, given that the situations are not the same. It is therefore necessary to stop the transposition in due time. This constitutes an important risk. The analogical explanation may suggest that the situation to be explained possesses some features that it does not have. For instance, for a series circuit, the analogy between electric potential and gravitational potential of a river may suggest that a given point in the circuit will always be at the same potential, which is erroneous. On this basis, we may also think that changing something *downstream* has no effect *upstream*, which is no less wrong. Yet, it is noteworthy that many words used in physics are steeped in metaphors. Regarding this, Ogborn (1996) reminds us that Faraday invented the words "ion" and "electrode" which, via their Greek origin, respectively evoke "an object that goes" and "a way for electricity" (p. 74).

We might note that the anthropomorphism often observed in popular science-like explanations is just a particularly hazardous form of analogy. Finalist explanations can even be seen as sharing a form of anthropomorphism: regarding osmosis for instance, it is often said that water crosses a semi-permeable membrane from the less concentrated towards the more concentrated solution, *in order to* lower the concentration of this latter solution, that is, *as if it wanted to* (Appendix I). As a rule, metaphors and analogies can be very useful for explaining physical phenomena, but a vigilant critical analyst will carefully consider the limits of these tools. It would be naive to let us be lulled into metaphorical poetry. We already have many problems with the "black" of the "hidden mass" of the universe, that of black bodies or that (the same?) of "holes" that are no less "black".

Given its explicit character, analogy permits, more than metaphor, a detailed and controlled parallelism between the phenomenon to be explained (the *target*) and the situation that should aid its comprehension (the *source*). It highlights the similarity of two formal structures which might account for the two situations being compared. Several examples below will show the benefits that can be expected from a cautious use of analogies (Appendices D, F and G).

3.5 Multiple Risks: Focal Points to Guide Critical Analysis

When considering the dangers to which explanatory lines of reasoning are exposed, we are struck by their multiplicity. It is also noteworthy that, for a given text, several of these risks may be simultaneously detected. The document about Young's slits (Fig. 3.14) is at the same time a case of similarity of symbols and of hyper-selectivity of the represented elements. The same can be said of the document about colour syntheses (Fig. 3.15). Similarly, a text may present both a story-like structure and a somewhat risky metaphorical language, such as the analysis of the change of state discussed above ("molecules cannot resist any longer..."), these two features being supposed to captivate the reader. That said, those risk factors will not necessarily put strong limits on the explanatory value of a text. To appreciate this point, it is essential to consider the targeted audience. A child may well be misled by the image showing some light rays behind a magnifying glass; this will be less frequently the case for adults well-read in science. Moreover, all these risky features aim at the readers' (or listeners') intellectual comfort, therefore their utility should be judged in terms of compromise.

From the perspective of an education to critical analysis, we discuss in the following chapters how the elements of analysis characterized in this chapter can contribute to guiding a critical study of text.

References

Bachelard, G. (1938). *La formation de l'esprit scientifique*. Paris: Vrin.

Bächtold. (2014). L'équation $E_{libérée}=|\Delta m|c^2$ dans le programme et les manuels de première S. *Recherche en Didactique des Sciences et des Techniques, 10*, 93–122.

Besson, U. (2004). Students' conceptions of fluids. *International Journal of Science Education, 26*(14), 1683–1714.

Boizier, C. (2012). Difficultés associées à l'usage d'un texte historique de première main : Travail sur la description de l'expérience démontrant la composition de l'air par Lavoisier en classe de quatrième. Master de didactique des disciplines. Université Paris Diderot-Paris 7.

Botinelli, L., Brahic, A., Gouguenheim, L., Ripert, J., & Sert, J. (1993). *La Terre et l'Univers*. Paris: Hachette.

Centre National de Documentation pédagogique (France). (2000). *Document d'accompagnement du programme de Seconde Générale* (grade 10).

Colin, P., & Viennot, L. (2001). Using two models in optics: Students' difficulties and suggestions for teaching, Physics education research. *American Journal of Physics Sup., 69*(7), S36–S44.

Colin, P., Chauvet, F., & Viennot, L. (2002). Reading images in optics: Students' difficulties, and teachers' views. *International Journal of Science Education, 24*(3), 313–332.

Crawford, F. S. (1965). *Berkeley physics course vol. 3, waves*. New York: McGraw-Hill.

de Hosson, C., & Kaminski, W. (2007). Historical controversy as an educational tool: Evaluating elements of a teaching–learning sequence conducted with the text "dialogue on the ways that vision operates". *International Journal of Science Education, 29*(5), 617–642.

Driver, R., Guesne, E., & Tiberghien, A. (Eds.). (1985). *Children's ideas in science*. Milton Keynes: Open University Press.

Euler, M. (2004). The role of experiments in the teaching and learning of physics. In E. F. Reddish & M. Vicentini (Eds.), *Research on Physics education* (International school of physics Enrico Fermi). Amsterdam: IOS Press.

Fauconnet, S. (1981). Etude de résolution de problèmes: quelques problèmes de même structure en physique, Thèse de troisième bicycle, Université Paris 7.

Feller, I., Colin, P., & Viennot, L. (2009). Critical analysis of popularisation documents in the physics classroom. An action-research in grade 10. *Problems of education in the 21st century, 17*(17), 72–96.

Gauvrit, N. (2007). *Statistiques méfiez-vous!* Paris: Ellipse.

Gentner, D. (1983). Structure-mapping: A theoretical framework for analogy. *Cognitive Science, 7*(2), 155–170.

Halbwachs, F. (1971). Réflexions sur la causalité physique. In M. Bunge, F. Halbwachs, T. S. Kuhn, J. Piaget, & L. Rosenfeld (Eds.), *Les Théories de la Causalité*. Paris: Presses Universitaires de France.

Jacobi, D. (2005). *Les sciences expérimentales communiquées aux enfants*. Grenoble: Presses Universitaires de Grenoble.

Kahneman, D. (2012). *Thinking fast and slow*. London: Penguin books.

Kaminski. (1989). Conception des enfants et des autres sur la lumière. *Bulletin de l'Union des Physiciens, 716*, 973–996.

Karplus, R. (1969). *Introductory Physics a model approach*. New York: Benjamin Inc.

Kress, G., & Van Leeuwen, T. (1996). *Reading images: The grammar of visual design*. London: Routledge & Kegan Paul.

Lavoisier, A. L. (1789). *Traité élémentaire de chimie* (Vol. 1, planche IV, Fig. 2).

Lessons by Marie Curie, collected by Isabelle Chavannes in 1907. (2003). *Physique élémentaire pour les enfants de nos amis*. Coord. B. Leclercq. Paris : EDP Sciences.

Maury, J.-P. (1989). *La glace et la vapeur, qu'est-ce que c'est ?* Paris: Palais de la Découverte.

Maury, L., Saltiel, E., & Viennot, L. (1977). Etude de la notion de mouvement chez l'enfant à partir des changements de référentiels. *Revue Française de Pédagogie, 40*, 15–25.

McLelland, J. A. G. (2011). A very persistent mistake. *Physics Education, 46*(4), 469–471.

Michaut, C. (2014). *La vulgarisation scientifique : mode d'emploi*. Les Ulis: EDP Sciences.

Ogborn, J. (1996). *Explaining Science in the classroom* (pp. 70–71). Buckingham: Open University Press.

Pinto, R., Ametler, J., Chauvet, F., Colin, P., Giberti, G., Monroy, G., Ogborn, J., Ormerod, F., Sassi, E., Stylianidou, F., Testa, I., & Viennot, L. (2000). Investigation on the difficulties in teaching and learning graphic representations and on their use in science classrooms, *STTIS transversal report (WP2)*, www.uab.es/sttis.htm and http://www.lar.univ-paris-diderot.fr/sttis_p7/index.htm

Reflets de la physique. (2017). n° 52, Paris : Société Française de Physique.

Rozier, S. (1988). Le raisonnement linéaire causal en thermodynamique classique élémentaire. Paris, Thèse, Université Paris 7.

Rozier, S., & Viennot, L. (1991). Students' reasoning in thermodynamics. *International Journal of Science Education, 13*(2), 159–170.

Valentin, L. (1983). *L'univers mécanique*. Paris: Hermann.

Viennot, L. (2001). *Reasoning in physics the part of common sense*. Dordrecht: Springer.

Viennot, L. (2004). Common reasoning in physics : Relations fonctionnelles, chronologie et causalité, In L. Viennot, & C. Debru (dir.) *Enquête sur le concept de causalité* (pp. 7–29). Paris: PUF.

Viennot, L. (2007). La physique dans la culture scientifique: entre raisonnement, récit et rituels, *Aster* n° spécial « Science et récit », n°44, 23–40.

Viennot, L. (2014). *Thinking in physics, the pleasure of reasoning and understanding*. Dordrecht: Springer.

Viennot, L., & De Hosson, C. (2012). Beyond a dichotomic approach, the case of colour phenomena. *International Journal of Science Education, 34*(9), 1315–1336. https://doi.org/10.1080/09500693.2012.679034.

Viennot, L., & De Hosson, C. (2015). From a subtractive to multiplicative approach, a concept-driven interactive pathway on the selective absorption of light. *International Journal of Science Education, 37*(1), 1–30. https://doi.org/10.1080/09500693.2014.950186.

Chapter 4
Benefits and Limits of a Classification

4.1 Implementing a Critical Analysis: First Approach

Following the discussion of types of flaws and risk factors in texts in Chaps. 2 and 3 (summarized in Tables 4.1 and 4.2), we will now discuss how we might approach an explanatory text using the framework that emerges from this.

4.1.1 A Document on Osmosis: A Case Where the Flaws Are Relatively Obvious

Let us now imagine a document consisting of a drawing (Fig. 4.1) and a text about osmosis.

The text comments a figure analogous to Fig. 4.1 as follows (Bouissy et al. 1987, p. 110):

> Since everything happens as if the membrane were not-existent for the solvent, a situation of non-equilibrium will occur (…) the flow of solvent continues until a new equilibrium is realised (minimum free energy). Therefore, when the concentrations in A and B become equal, there will be different levels in the two compartments (…), and therefore a difference in pressure $\Delta p = p_A - p_B$, (…) called osmotic pressure.

This composite document illustrates several types of flaws.

First, there is an indirect contradiction between this paper and the accepted theory of osmosis. Indeed, at the equilibrium following the initial situation, the concentrations of solute in the two compartments are not identical. At equilibrium, the "chemical potential of the solvent in the presence of solute", a function of absolute temperature T, pressure p and molar concentration C, is the same on both sides of the membrane (for dilute solutions of small molecules, this function takes the following expression: $\mu(T, p, C) = \mu_0(T, p) - CkT$, where $\mu_0(T, p)$ is the chemical

L. Viennot, N. Décamp, *Developing Critical Thinking in Physics*, Contributions from Science Education Research 7,
https://doi.org/10.1007/978-3-030-43773-2_4

Table 4.1 Summary of the main types of flaws in critical text analysis

Types of flaws	Description in this book
Internal contradiction	p. 11
Direct contradiction of a law	p. 13
Indirect contradiction of a law	p. 13
Logical incompleteness	p. 17
Over-generalisation	p. 18
Incompatibility with a thought experiment	p. 20

Table 4.2 Summary of the main risk factors for critical text analysis

Risk factors	Description in this book
Exactness of the text's conclusion	p. 27
Echo-explanation	p. 30
Non-standard designation of the entities concerned	p. 30
All or nothing	p. 31
"Small" assimilated to "zero"	p. 32
Only one cause taken into account	p. 33
Only one location taken into account	p. 35
Explicit story-like explanation	p. 37
Suggested story-like explanation	p. 40
Image: Realism and symbolism	p. 46
Image: Similarity of symbols	p. 47
Image: Overselectivity	p. 48
Image: Structure and scales	p. 51
Analogy/metaphor	p. 53

potential of the pure solvent and k is the Boltzmann constant, see, e.g., Diu et al. 1989, (p. 408). It follows that, at equilibrium, if the concentration and temperature are the same on both sides of the membrane, the pressure is the same. However, a reader of the text may not be familiar with the theory in question.

A second flaw is revealed by a thought experiment: let's imagine that pure water is put into the left-hand compartment. It would pass continually into the other compartment, where the concentration of solute would always be higher. This does not happen: you don't need an experiment to convince yourself of this (even if it's interesting to try it, as in Fig. 4.2).

Finally, a third flaw is revealed by another "thought experiment": In the final state assumed by the author, the levels are different and the concentrations are identical. But it is possible to imagine a situation (Fig. 4.3a) where the content of the two branches is identical to the final content of the left branch shown in Fig. 4.1; everything is then identical on both sides of the membrane, it is a balanced situation. Here are two possible equilibria (Fig. 4.3a, b), with the same mass of liquid on the left in equilibrium with two different masses of liquid on the right, all with the same concentration. There are good reasons to be surprised (Table 4.3).

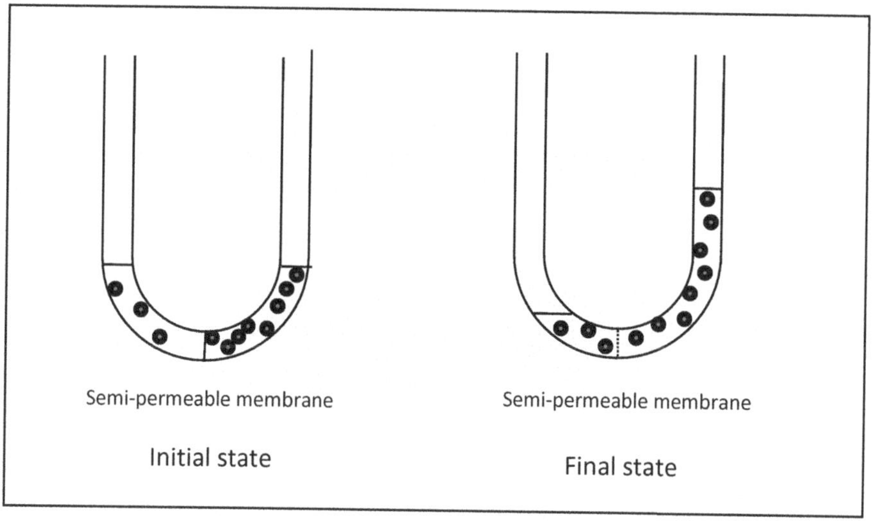

Fig. 4.1 A representation for the evolution from an initial state to a situation of osmotic equilibrium (equivalent diagrams are frequently observed, e.g. in Wikipedia in French and in English, see also Bouissy et al. p. 110)

How could such a document be constructed and accepted by its author and proofreaders? Factors that may have influenced the author are reported in Table 4.4.

This analysis leads to several remarks.

First, not all risks are equally consequential for all types of audiences. The fact that a small number of solute molecules appear in the form of clearly visible coloured circles scattered over a colourless continuous background representing water will probably not disturb a graduating student, whereas a child can misunderstand such an image.

Secondly, all the categories could be debated. For example: can we assert that the drawing in Fig. 4.1 suggests a story-like explanation simply because the final equilibrium, with same concentration in the two branches, seems to be determined by a single cause (initial difference in solute concentration)? That said, why worry about what is suggested when the same is explicitly expressed in the text?

Finally, we can answer with a *yes and no* concerning the coincidence between the content of the text and accepted physics. This text is inaccurate since it is not possible to have an equilibrium with the same concentration of solute and different pressures on either side of the membrane. But it respects some of the valid statements: the initial cause of the phenomenon is indeed a difference in the concentration of the solute, and the end state has a difference in pressure which is called osmotic pressure. These remarks underline that the value of such a classification is not as a formal exercise. The proposed categories of flaws are there only to stimulate our critical reflection.

Once an analysis by category of possible flaws has been carried out, an overall judgment can be expressed. Here, it is probably the uniqueness of the cause of a

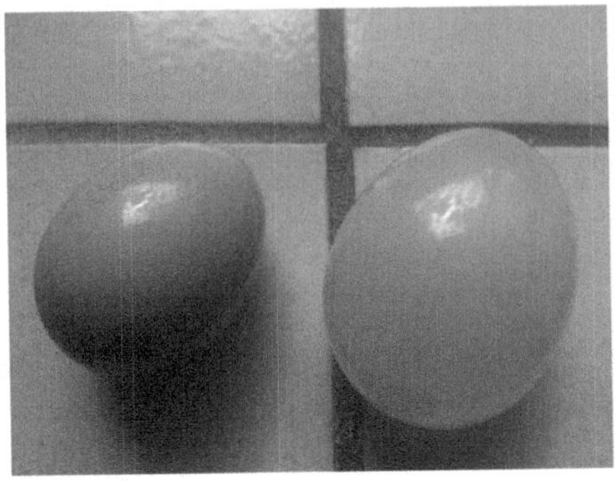

A raw egg stripped of its shell by prolonged soaking in vinegar is separated from the outside by a semi-permeable membrane. Then soaked in pure water, its volume does not increase indefinitely. Left: an egg with the initial size of the egg shown after osmosis (right).	At equilibrium, the semi-permeable membrane is under tension. During soaking, the pressure in the egg increased and the molar concentration of solute decreased. At equilibrium, these quantities are both different from their respective values in pure water.

Fig. 4.2 A domestic realization of an osmotic equilibrium between a solution (inside the egg) and pure water: at equilibrium, the transfer of pure water is limited by the internal overpressure of the egg

phenomenon without reference to feedback from the effect produced which is the main characteristic of this text and leads to error. It turns out that, despite the complexity of the theme, the detection of this error is relatively easy thanks to the thought experiments described above.

4.1.2 Capillary Rise: Not So Simple

Let's now imagine a document on capillary rise in a tube. It consists of a diagram (Fig. 4.4) and a text.

This text is brief (Quéré 2001): "The force that lifts the liquid is $\gamma_{SG}-\gamma_{SL}$". The liquid stops rising when the force that lifts it is balanced by the weight." (p. 158).

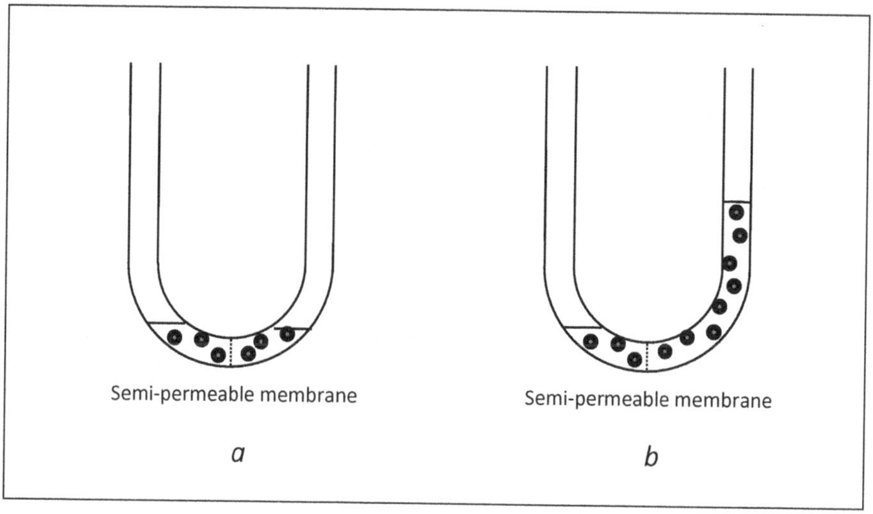

Fig. 4.3 Two situations of equilibrium both possible according to the text under examination. One (**a**) is totally symmetric and is obviously at equilibrium and the other (**b**) is like in Fig. 4.1: the same content on the left hand side would be in equilibrium with two very different contents on the right hand side

Table 4.3 Types of flaws present in the document *Osmosis*

Internal contradiction	–
Direct contradiction of a law	–
Indirect contradiction of a law	Yes
Logical incompleteness	–
Over-generalization	–
Incompatibility with a thought experiment	Yes, with two thought experiments

Tables 4.5 and 4.6 summarize the types of flaws and risk factors that can be identified in this document.

In terms of classification, the same remarks can be made as in the previous example: variable decisiveness depending on the items, relativity of judgments to the intended audience. But we are in a less simple case. A dominant aspect emerges, in this case the localization of the supposed forces acting at the top of the column, which goes with the expression: "The force that lifts the liquid". However, to identify the flaws in the underlying reasoning, a complex analysis is needed (see Appendix G). This is all the less obvious as the end of the story seems right: yes, the water is up in the tube; yes, Young's relationship gives the angle of contact. The inconsistent elements are not obvious, except for the unbalance of forces in horizontal direction, and for a difficulty in understanding what can balance the weight of the water column. There is no simple thought experiment to question the expression "The force that lifts the liquid". This explains why this type of presentation has enjoyed both popularity and longevity. But some authors (Berry 1971; Das et al.

Table 4.4 Types of risk factor related to the document *Osmosis*

Non-standard designation of the entities concerned	–
Accuracy of the text's conclusion	No, because at equilibrium the concentrations of solute in the two compartments cannot be identical. For solute concentrations to be equal at equilibrium, pressures on both sides of the membrane must be equal
	But there are elements that conform to the correct physics: There will be (…) a level difference between the compartments (…), so a pressure difference $\Delta p = pA - pB$, (…) called osmotic pressure. These elements can help to legitimize the text
Echo-explanation	Yes, by focusing on a single quantity (solute concentration) and using the narrative explanation. An incorrect result is arrived at by imagining that equilibrium is reached when the only cause envisaged disappears
All or nothing	–
"Small" assimilated to "zero"	–
Only one cause taken into account	Yes: The difference in solute concentration is the only cause taken into account. The counter-effect of the pressure difference is not considered
Only one location taken into account	–
Explicit story-like explanation	Yes: "(…) a situation of non-equilibrium will occur (…) the flow of solvent continues until a new equilibrium is realized (…). Therefore, when the concentrations in A and B become equal, there will be different levels in the two compartments (…), and therefore a difference in pressure $\Delta p = p_A - p_B$, (…)"
Suggested story-like explanation	Here the narrative structure of the explanation is explicit
Image: Realism and symbolism	One could criticize the representation of the solute, with (necessarily) out of scale molecules in small numbers, while water is not shown similarly as particulate. This kind of realism can seriously mislead only people very inexperienced in science, which is not so common among those interested in osmosis
Image: Similarity of symbols	
Image: Hyperselectivity	
Image: Structure and scales	
Analogy/metaphore	–

2011; Marchand et al. 2011; Planinsic 2016) emphasised the flaws in this type of text, particularly this one: a flat or cylindrical vertical glass wall can only attract a molecule of liquid horizontally, for symmetry reasons. It is therefore excluded that a force due to glass "lifts the liquid" by pulling on its upper surface. These authors have made coherent counter-proposals for the analysis of the phenomenon, which are detailed in Appendix G. Reading their suggestions, it seems clear that the culture of a physicist is needed to conduct such discussions. In this case, the category "indirect contradiction of a law" appears to be very broad to guide criticism on its own. Our first categories may be useful, but they do not by far exhaust the subject.

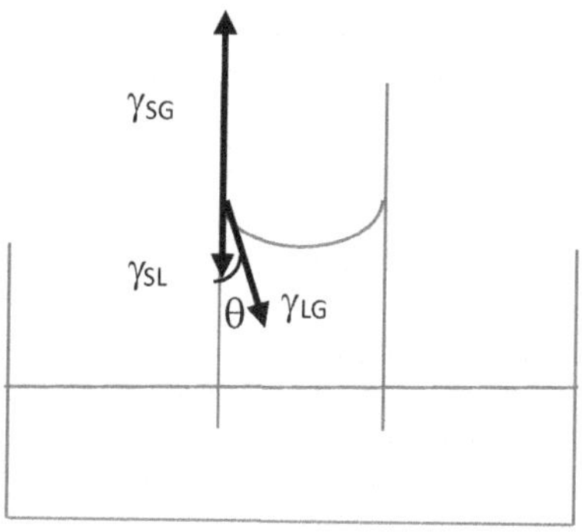

Fig. 4.4 A commonly used diagram to introduce Young's formula $\gamma_{LG}\cos\theta = \gamma_{SG} - \gamma_{SL}$, where γ_{LG}, γ_{SG} and γ_{SL} are interfacial coefficients (forces per unit length of triple contact line) for the liquid/gas, solid/gas and solid/liquid interfaces respectively, and θ is the contact angle

Table 4.5 Types of flaws related to the document *Capillary ascension*

Internal inconsistency	–
Direct contradiction of a law	–
Indirect contradiction of a law	Yes, if the three arrows are to be understood as a free body diagram: This diagram does not show forces that are balanced in the horizontal direction
Logical incompleteness	Difficulty seeing what balances column weight
Overgeneralisation	–
Incompatibility with a thought experiment	–

4.1.3 *Beyond the Classification*

These two examples suggest a first remark. When analysing a text critically, it may be useful to know a priori the types of reasons for rejection (Chap. 2) or doubt (Chap. 3) about its content: This guides the search for possible objections and raises useful questions. But it is probably illusory to expect to find some kind of flaw detection algorithm. The first example (osmosis) suggests that, no matter how complex a phenomenon may be, a relatively naïve critical look can detect that something is wrong with the proposed document. The second (capillary rise) leads us to recognize that this judgment is context-dependent. Generally speaking, the complexity of the content addressed in a text and the conceptual mastery of the person reading it should clearly be taken into account when activating (for oneself) or promoting (in others) a critical attitude. From here on, this book deals with this complex relationship – *a priori* – between critical attitude and conceptual mastery.

Table 4.6 Types of risk factor related to the document *Capillary ascension*

Non-standard designation of the entities concerned	Not in an obvious way. But it is suggested that the arrows correspond to forces (per unit of interface line length) acting on a line, i. e. on an element without mass. This prohibits a Newtonian budget type meaning
Accuracy of the text's conclusion	Yes, a water column rises above the free surface of the vessel, and the angle of contact is well defined by Young's relationship
Echo-explanation	Yes, with regard to the location of forces only at the top of the liquid column
All or nothing	–
"Small" assimilated to "zero"	–
Only one cause taken into account	–
Only one location taken into account	Yes, at the top of the column
Explicit story-like explanation	–
Suggested story-like explanation	–
Image: Realism and symbolism	Not harmful
Image: Similarity of symbols	Not harmful (although interface tensions and forces are represented in the same way)
Image: Overselectivity	Yes, since we would have to represent forces acting at places other than the top of the column, and we would have to specify the objects on which the forces act, in order to understand which objects are the subject of a force balance
Image: Structure and scales	Not harmful
Analogy/metaphor	Yes: A force *lifts* the liquid

References

Berry, M. V. (1971). The molecular mechanism of surface tension. *Physics Education, 6*, 79–84.

Bouissy, A., Davier, M., & Gaty, B. (1987). *Physique pour les sciences de la vie* (tome) (Vol. 2). Paris: Belin.

Das, S., Marchand, A., Andreotti, B., & Snoeijer, J. H. (2011). Elastic deformation due to tangential capillary forces. *Physics of Fluids, 23*, 072006.

Diu, B., Roulet, B., Lederer, D., & Guthmann, C. (1989). *Mécanique Statistique* (pp. 408–410). Paris: Hermann.

Marchand, A., Weijs, J. H., Snoeijer, J. H., & Andreotti, B. (2011). Why is surface tension parallel to the interface. *American Journal of Physics, 79*, 999–1008.

Planinsic, G. (2016). On the surface of it there is more to it. *Physics Education, 51*(2016), 030105.

Quéré, D. (2001). Bulles gouttes et perles liquides. In D. Jasmin, J. M. Bouchard, & P. Léna (Eds.), *Graines de science* (Vol. 3, pp. 145–165). Paris: Le Pommier.

Chapter 5
Conceptual Mastery and Critical Attitude: Complex Links

Is critical thinking an ability in itself that everyone can implement regardless of the content involved? This issue has been debated for ages. To speak of critical thinking, without further specification, would be to respond positively. For our part, in the absence of an academic consensus, we leave the door wide open to Willingham's (2007) previously quoted viewpoint: *Critical thinking is not a set of skills that can be deployed at any time, in any context* (p. 10).

The following reflections are based on research on physics education. They are intended to fuel the debate we have just mentioned, but above all to provide elements for the learning of critical analysis, whether it is to evolve personally or to train physics students or teachers. Note that when it comes to teaching physics, the stakes are high. Indeed, there is unanimity on the importance of developing critical thinking among students. In these times of an abundance of information on the web, we can clearly see the need for a critical filter. Moreover, it may seem tempting to make students work on documents like those that appear on the web on the grounds that they are presumably motivating. Under these conditions, it can become difficult to preserve the role of conceptual knowledge. The debate about the idea of *critical thinking* 'in itself' is then formulated in the following terms: can one train for critical analysis without the support of minimal conceptual knowledge of the field in question?

In addition to understanding, another dimension seems essential in clarifying this issue. In fact, to put it briefly, it is a question of multiple dimensions which might be termed psycho-cognitive. Indeed, adopting a critical attitude towards an explanation in physics implies a certain idea of what learning in science is, and also a lucidity about one's own state of understanding (e.g. Vermunt 1996). Underlying any critical response is intellectual frustration or more positively, a search for intellectual satisfaction (Viennot 2006; Mathé and Viennot 2009). It also means feeling entitled to understand, an attitude that goes along with a certain self-esteem (Bandura 2001).

L. Viennot, N. Décamp, *Developing Critical Thinking in Physics*,
Contributions from Science Education Research 7,
https://doi.org/10.1007/978-3-030-43773-2_5

Separating these different metacognitive and affective factors is not easy, but at least it can be argued that they are very likely to intervene in the intellectual evolution of a person who reads a text and seeks both to learn and to evaluate it critically.

5.1 Future Teachers Respond to Radio-Carbon Dating Explanations

Research results (Décamp and Viennot 2015) are now briefly summarised. They illustrate different types of intellectual dynamics that can be observed during an interview.

5.1.1 Interviews On a Theme that Is Not As Simple As It Sounds

Ten teachers at the end of their teacher education were interviewed individually for 70 to 90 minutes. The subject was carbon-14 dating, already referred to in Chap. 2. This technique is based on the law of radioactive decay giving the number N(t) of radiocarbon atoms (^{14}C) of an initial population N_0 remaining after a time t: $N(t) = N_0 \exp(-\lambda t)$. The value of the constant λ is simply linked to the time after which only half of the initial atoms remain ("half life", in this case 5730 years) (see Appendix C for more detail). Beginning teachers were asked whether or not they were satisfied with various explanations. Six texts were presented to them in turn, each time with the same questions: is this text appropriate for you and why? Would you like something to be changed or added and why? As the interview progressed, the texts presented provided more and more complete, or rather less and less incomplete, explanations. The first is far from permitting an understanding of this technique, for example, it does not indicate how to find the $^{14}C/^{12}C$ proportion of the organism at the time of its death, which is essential for performing a dating at the time of discovery. The third text does not explain why the $^{14}C/^{12}C$ proportion in the atmosphere would be stable since the death of the organism, and the fifth text still does not explain why the temporal rates of formation and disintegration of ^{14}C in the atmosphere are equal. On the other hand, the last text, prepared by the authors of the study and based on the model presented in Appendix C, allows a simplified but self-coherent understanding of the phenomenon. During the course of the interview and at the request of the interviewees, information is provided to them (for example the mechanism of formation of ^{14}C from nitrogen, ^{14}N, which involves neutrons themselves stemming from atoms of molecules in the upper atmosphere impacted by cosmic rays).

5.1.2 Analysis of Interviews: Broad Outlines

In the interview transcripts, expressions of agreement ("Yes, I think this explanation is fine"), mixed agreement ("Yes, basically that sums it up well...") or disagreement were identified.

The questions asked by the beginning teachers interviewed were also coded. Their purpose can be directly related to the conceptual elements essential to understanding (the reader does not need to understand these points to understand what follows), i. e.:

- The need to know the initial proportion of radiocarbon to ordinary carbon [$^{14}C/^{12}C$] in an organism at the time of its death;
- The fact that it was then the same as in the atmosphere.
- The hypothesis that the ratio [$^{14}C/^{12}C$] in atmosphere has not changed since the death of the organism;
- The process of formation of radiocarbon;
- The process of radioactive decay of radiocarbon;
- How the balance between the corresponding numbers per second of radiocarbon atoms involved in these two processes results in a steady value of [$^{14}C/^{12}C$] in the atmosphere;
- The constancy in time of the total number of nuclei (radio carbon + nitrogen);
- The multiplicative effect of the existing numbers of radiocarbon and nitrogen nuclei in the destruction and formation of ^{14}C nuclei, respectively;
- The hypothesis that the ratio [$^{14}C/^{12}C$] in atmosphere has not changed since the death of the organism;
- How this multiplicative structure explains the stable proportion of radiocarbon to ordinary carbon in the atmosphere.

These interventions were then coded as a "crucial question" (**cq**). In the opposite case, interviewees' questions are coded as a "question about a detail" (**dl**). For example: "Why are these two rates equal, by chance?" was coded as "crucial" and: "How does the detector work?" was coded as "question about a detail". As regards psycho-cognitive aspects, expressions of satisfaction or frustration were also identified over the course of the interview, while an overall evaluation by beginning teachers at the end of the interview indicated their final feeling of satisfaction with a number from 1 (low satisfaction) to 4 (high satisfaction). Table 5.1 shows the codes corresponding to the types of questions asked and levels of agreement, text by text, while Table 5.2 summarises the expressions of satisfaction or dissatisfaction of participants at each step, that is, at the end of the discussion about each text. In both tables, participants are ranked in order of their first crucial question.

Table 5.1 Level of agreement at the end of each step (S) and type of question asked

Participant	S_1	S_2	S_3	S_4	S_5	S_6	**Overall satisfaction**
G	≈ **dl**	≈ **cq**	**cq**	**cq**	**cq**	Θ	3
B	Θ	Θ **cq**	**cq**	**cq**	**cq**	Θ	4
S	≈ **dl**	≈ **dl**	≈	**cq**		Θ	3
J	Θ	Θ	Θ **dl**	**cq**	**cq**	Θ	3
M	Θ	Θ	Θ	**dl cq**	**cq**	Θ	3
T	≈ **dl**	≈ **dl**	Θ	**cq**	≈	Θ	2
A	≈	Θ	Θ		**cq**	Θ	2,5
V	Θ	Θ	Θ **dl**	≈	**cq**	Θ	3
Y	≈ **dl**	Θ **dl**	Θ	Θ	Θ	Θ	4
H	Θ	Θ **dl**	Θ **dl**^{2}	Θ	Θ	Θ	4

Notations: Θ: agreement; ≈: mixed agreement; dl: question about a detail; cq and shaded box: crucial question

5.1.3 Some Striking Facts

Apart from the last two participants (Y and H), the same diagonal (see the shaded areas) divides Tables 5.1 and 5.2 into two parts. Unsurprisingly, we find only questions of detail to the left of this diagonal since Table 5.1 is built on this rule. It is more significant that, to the left of this same diagonal, what dominates is the level of agreement (albeit sometimes qualified) that the participants express with regard to the texts presented to them, however incomplete.

(Step 1) Participant H – Well, that's enough for me, but then, if you ask me more specific questions, I might remember something else…

(S1) J – No, that's enough, well, that's pretty clear too. We understand the principle, anyway.

(S1) V – Uh, he's presenting the main points, so maybe, after maybe, I don't know, wait for me to reread some of the parts, maybe some of them would need more detail. [reading the text] Hmm, No, to me, yes, he gives the gist….

(S1) M – Yeah, no no, it's a good starting point. So it's true that it's very succinct, but we can't do a whole course either… Hmm. No, no, it's clear! Clear, concise …

(S3) Investigator – Is it always clear? Is it always enough, do you lack any information or not?

Table 5.2 Level of agreement at the end of each step (S) and statements expressing satisfaction or frustration

Participant	S_1	S_2	S_3	S_4	S_5	S_6	**Overall satisfaction**
G	≈	≈ *m-*	*m-*	*m-*	*m-*	Θ *m+*	3
B	Θ	Θ *m+*	*m+ m-*	*m-*	*m-*	Θ *m+*	4
S	≈	≈	≈	*m-*	*m-*	Θ *m+*	3
J	Θ	Θ *m+*	Θ	*m-*	*m-*	Θ *m+*	3
M	Θ	Θ *m+*	Θ	*m-*	*m-*	Θ *m+*	3
T	≈ *m-*	≈ *m-*	Θ		≈ *m-*	Θ *m+*	2
A	≈	Θ *m+*	Θ *m+*	*m-*	*m-*	Θ *m+*	2,5
V	Θ	Θ	Θ	≈	*m-*	Θ	3
Y	≈	Θ *m+*	Θ *m+*	Θ	Θ	Θ *m+*	4
H	Θ	Θ	Θ	Θ	Θ	Θ	4

Notations: Θ: agreement, ≈: mixed agreement; *m+*: satisfaction after providing information; m^- and shaded box: frustration

T – Um, no, I'm fine. That's enough for me.

With regard to their comments expressing satisfaction or frustration (Table 5.2), those mentioned to the left of the shaded area say that a new text brings more information, but without any real criticism of the previous text:

(Step 2) Participant A – I think that's what I was missing earlier to know exactly how it was going. I think I forgot.

(S2) B -Yes well, this one seems more precise to me. Finally…

(S3) A – This one is much more precise and more specific, I'm not sure, well, there is not necessarily a need to detail so much to understand the principle. Then it depends on how demanding you are.

(S2) M – Finally, I think that the two texts complement each other well. It's true that a brief explanation of where the ^{14}C comes from is very quick, as in the first text, it's sympathetic, and afterwards how we exploit it in dating, in the second text, it completes well I find.

On the contrary, to the right of the diagonal, in the zone grayed out in Table 5.2, there is no longer agreement, and it is frustration that dominates, punctuated by crucial questions:

(S4) B – I have more of a problem with that than what it explains.

(S5) Interviewer – In fact, it's no coincidence. What this text says is that it is not a coincidence, it is because a stationary state has been established.
A – Well, that doesn't really explain that …
Interviewer – Doesn't really explain.
A – It's not an explanation! (…) No, no, it's true, I have more questions than when I arrived.
(S4) Interviewer – Hum hum. So, is it clear, is it enough, is there something missing?
J – Doesn't explain to me why… (…) Why it's equal (*these are the temporal rates of formation and decay of radiocarbon*)?

This passage from docile approval to a clear expression of their frustration is sometimes accompanied by an awareness of their previous attitude, so this ironic comment by a participant on her previous answers:

(S4) Interviewer – Do you think it's clear, too? Do you miss something?
A – Well, that sounds like a yes [*laughter*], since I didn't understand anything!

5.1.4 Delayed Critique: An Activation Threshold

To summarise, most participants showed threshold behaviour. At a certain point in the discussion, they took ownership of the problem, distanced themselves from the texts discussed, realised that they need a more complete explanation, asked crucial questions and argued more strongly. One might think that this trigger occurs at some stage of their conceptual progression. They may still be far from a complete understanding of the subject, but they are beginning to appreciate its complexity and crucial aspects of it. This threshold of understanding, which varies from one participant to another, determines when they feel the need and dare to express their frustration. Figure 5.1 illustrates graphically this hypothetical "delayed critique" model.

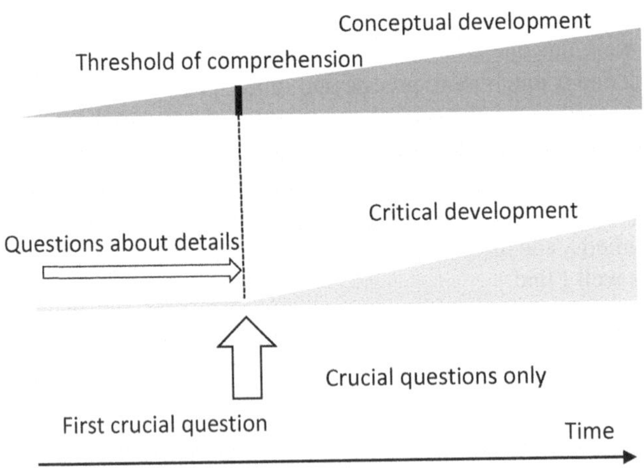

Fig. 5.1 A model of intellectual dynamics – a *delayed critique*

The most remarkable aspect of this study is the prolonged a-critical phase that marks the first contact of a person with an explanation that he or she has all the means to challenge.

5.1.5 *Expert Anesthesia*

Two individuals in the study sample differed quite noticeably from this description. Throughout the interview, they did not express any criticism of the texts presented. These beginning teachers were very familiar with the subject and did not ask any crucial questions. At the end of the interview, they said they were very satisfied, but it was above all because they answered correctly ("I am very happy with my answers"). It should be noted that they were as inactive in critical analysis of the documents reviewed as they were competent regarding radiocarbon dating.

5.2 Delayed Critique or Expert Anesthesia: A Fatality?

The main finding of this radiocarbon study is that, in order to activate their critical potential, most of the participants needed to reach a threshold of understanding. Once this threshold is reached, more or less quickly depending on the individual, an attitude of agreement, implicit satisfaction, and questioning on incidental points turns towards an expression of frustration, crucial questions, criticism and self-criticism. The search for school reminiscences gives way to an urgent search for understanding. This triggering of the critical attitude evokes a change in phase, as physicists say, from a conciliatory attitude towards texts that are very incomplete, to a "critical crisis" that excludes any agreement and delays any expression of intellectual satisfaction. This type of intellectual dynamics, "a delayed critique", occurs for the majority of participants in this study.

As regards the threshold of comprehension, it could be argued that it is not surprising that it is necessary to understand what we are talking about before making critical judgements on a text. But what is not obvious is that there is no logical need to understand the whole phenomenon in question. If the radiocarbon composition of the atmosphere is assumed to be constant over time, the question that should be posed is this: is there no radioactive decay in the atmosphere? Some would call this question "naïve". This kind of naivety is in fact a highly valuable training objective: not to be afraid of your own ignorance and to demand a minimum of consistency in the explanations you receive. In this regard, the links that we observe between the development of conceptual understanding and the activation of a critical attitude must be taken into account, but are not (yet) clear.

These links even take a paradoxical form when sustained absence of critique is observed in beginning teachers who are already proficient in the subject, however incomplete the explanation analysed, These experts (for this subject), satisfied with their own answers, could therefore neglect to look at the texts for what they were. They may have subconsciously completed what they read there. This "expert anaesthesia" syndrome is found in a minority here, but our sample was not very exposed to it because the subject, which was relatively difficult, was initially not mastered by most of the participants in our study.

After this exploratory research, the questions that arose were multiple. On the one hand, would these results be similar in other fields of physics? On the other hand, is it possible to avoid these two intellectual dynamics, that of delayed critique due to insufficient conceptual mastery or that of critical anesthesia due to conceptual expertise? The first of these questions is explored in an earlier study of 14 undergraduate students in journalism and science communication (Mathé and Viennot 2009). The text submitted to them was an alleged popular science article on the theme of the isobaric balloon (Chap. 3). With one exception, apprentice journalists who were interviewed individually expressed their criticism of the article long after realizing the inconsistency of the isobaric pressure hypothesis. It is therefore tempting, *a posteriori,* to interpret this attitude as delayed critique. In addition, the radiocarbon study inspired three other similar studies on the proper use of survival blankets, on osmosis and on capillary rise (for a synthesis see Viennot and Décamp 2018). Cases of "delayed critique" were the most frequent in all these studies. The population concerned (in the case of the hot air balloon) or the complexity of these subjects (survival blanket, osmosis, capillary rise) probably explain why we did not observe any case of expert anesthesia on these themes. On the other hand, rare cases of "early critique" confirm that it is possible to express a relevant critique when faced by a text without having an advanced mastery of the field in question (Table 5.3).

For us, developing this attitude seems to be a priority objective: a good reason to document it further.

Table 5.3 Three types of critical response when discussing a questionable explanation in physics (Viennot and Décamp 2018)

Type of critical response	Description	Observed frequency in beginning teachers
Expert anesthesia	Absence of criticism in a person who has all the necessary knowledge in a field to detect an inconsistency or serious inadequacy in an explanation relating to that field	Variable frequency according to the physical domains
Delayed critique	Case of a person who wants to know more about a field before making the slightest criticism of an explanation relating to this domaine, even if no specialized knowledge is required	Majority of the population
Early critique	Case of a person who expresses a relevant criticism even if she knows very little in the field concerned	Rare

References

Bandura, A. (2001). Social cognitive theory: An Agentic perspective. *Annual Review of Psychology, 52*(1), 1–26. https://doi.org/10.1146/annurev.psych.52.1.1.

Décamp, N., & Viennot, L. (2015). Co-development of conceptual understanding and critical attitude, Analysing texts on radio-carbon dating. *International Journal of Science Education, 37*(12), 2038–2063.

Mathé, S., & Viennot. (2009). Stressing the coherence of physics: Students journalists' and science mediators' reactions. *Problems of education in the 21st century. 11*, 104–128. Retrieved from http://journals.indexcopernicus.com/abstract.php?icid=886218

Vermunt, J. D. (1996). Metacognitive, cognitive and affective aspects of learning styles and strategies: A phenomenographic analysis. *Higher Education, 31*, 25–50. https://doi.org/10.1007/BF00129106.

Viennot, L. (2006). Teaching rituals and students' intellectual satisfaction. *Physics Education, 41*, 400–408.

Viennot, L., & Décamp, N. (2018). Activation of a critical attitude in prospective teachers: From research investigations to guidelines for teacher education. *Physical Review Physics Education Research, 14*, 010133. https://doi.org/10.1103/PhysRevPhysEducRes.14.010133.

Willingham, D. T. (2007). Critical thinking, Why is it so hard to teach? *American Educator, 31*, 8–19.

Chapter 6
Activating Criticism Without Delay

6.1 A Potential for Criticism

From the outset we avoided speaking of critical thinking as if it were a determining quality of an individual. We prefer to suggest a potential for criticism that each person more or less willingly activates in response to a given text, sooner or later. However, targeted studies of teachers at the end of their preparation suggest that the absence of critique is frequent even when it is not justified by any logical necessity. In other words, there is often a late activation of a critical potential that is present (since it eventually manifests itself), but that seems inhibited. Critical analysis becomes possible by learning to overcome this inhibition. It is encouraging to see how some future teachers achieve this.

6.2 Examples of Early Critique

6.2.1 Survival Blanket

Appendix H discusses how best to use a survival blanket to protect yourself from the cold (Viennot and Décamp 2016a). An ordinary survival blanket has a silver side (more reflective and less emissive) and a gold side (more emissive and less reflective). The interest of this theme is that it leads to a discussion of the compromise between two imperatives (Fig. 6.1): that of returning to oneself as much energy as possible (one thinks of putting the silver side towards oneself), and that of emitting as little energy as possible towards the outside (for the external side, it is also the silver side that, all else being equal, emits the least).

A very common idea is that you have to put the silver side inwards to send "heat" (i.e. energy) back to yourself. The instructions for use reflect this preconceived idea

L. Viennot, N. Décamp, *Developing Critical Thinking in Physics*,
Contributions from Science Education Research 7,
https://doi.org/10.1007/978-3-030-43773-2_6

Fig. 6.1 On which side should the survival blanket be placed to protect against the cold? A dilemma!

without reservation. But returning more energy to the inside means putting the golden side outwards and thus emitting more energy to the ambient air. Unless both sides of the cover are silvered, there is a dilemma. To doubt this, it is necessary to challenge a widely held misconception. As expected, the study participants – teachers at the end of their education – were slow to question the instructions for use that reflected the preconceived idea, especially since it coincided with their intuition. At the beginning of the interview, their memory had been refreshed as to the radiative properties of each face of the blanket. To this end, a simple experiment was performed: With a commercially available infrared radiometer, the students could compare the respective emissions of survival blankets stuck, with one side or the other outwards, on to a metal kettle full of boiling water (Appendix H). Note that resolving the dilemma is far from immediate. Not surprisingly, none of the participants was comfortable: no expert anesthesia here!

The finding that we stress here is that one of the interviewees was quickly able to question her initial idea of "heat" to be sent back to herself. She had just noticed the relatively large value indicated by the radiometer for the emission of the golden side towards the outside of the kettle. "I have a problem", she said soberly. Without knowing anything or almost nothing about the radiative properties of the various parts of the system involved- the body to be protected, the survival blanket, the ambient air – and even less about the energy transfers involved (radiative, conductive, convective), she had arrived at this doubt by trying to reconcile the little she knew: it is good to preserve the energy between herself and the blanket, but not to dissipate it outwards. This example of early critique is quite representative of what we might wish to develop in physics education.

6.2.2 Osmosis

Two other examples of early critique concern osmosis. It is on purpose that we still use a non-obvious topic: osmosis is not a simple subject. The study (Viennot and Décamp 2016b) summarized in Appendix I shows this clearly: as expected, no expert anesthesia was observed amongst future teachers.

Confronted with the document already discussed above (Fig. 4.1), which suggests an equal solute concentration on both sides of the membrane with different liquid heights at equilibrium, a beginning teacher was able to make this unambiguous comment:

> *Interviewee*- So to convince someone that it is not possible, this drawing, there, starting from this situation (*gestures indicating the same level*), I would add height in one of the compartments.
> *Interviewer* - Yes.
> - I'd say I just disturbed the balance.

The thought experiment suggested by this participant is judicious (p. 58): a situation where both liquid heights and concentrations are the same on both sides of the membrane is in equilibrium (Fig. 4.3). It seems therefore unlikely that a situation where the concentrations are the same but the liquid heights are different would also be at equilibrium (Fig. 6.2).

Another participant referred to the situation where initially pure water was on one side of the membrane, but not on the other:

> -I think it would not be possible, it won't be able to go up indefinitely, the leg (...) There is a balance of concentrations to respect but there is certainly also a difference in pressure not to be exceeded.

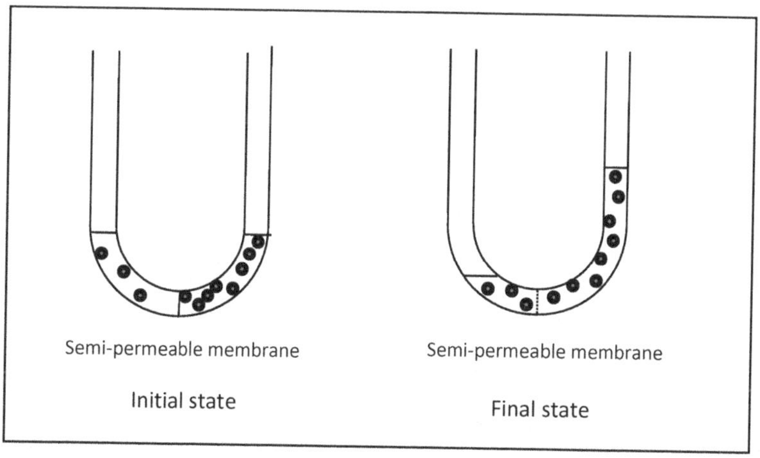

Fig. 6.2 Repeat of Fig. 4.1, a drawing frequently observed, notably on Wikipedia, to introduce the phenomenon of osmosis

The thought experiment invoked this time consists in taking the text at face value and examining an extreme case: starting with pure water on one side of the membrane. The water would then pass continuously to the other side where the concentration of solute would always be higher (as explained on p. 58).

6.2.3 Capillary Rise

A final example of early critique involves the very nature of the concepts involved. This is an interview with a teacher at the very beginning of her professional life. The discussion concerns the diagram in Fig. 6.3 already discussed in Chap. 4.

This diagram uses three surface tension coefficients in what resembles the balance of three forces acting on the contact line between three media: water, air and glass, except that they are forces per unit of line length. As shown in Appendix G, this diagram is open to several criticisms from the point of view of its explanatory value. In addition to not being balanced in a horizontal direction, it raises the question of its meaning, since it seems that the forces in question are exerted on an immaterial line. A participant in the research study stated, at the outset and several times thereafter, that she did not understand what these forces meant because it was unclear on which physical object they were exerted.

> Excuse me? What? A force on an interface? Well, no! Because all the molecules ... In fact an interface, it is the molecules that are, finally, … of one phase and another, they do not experience the same force.

This participant is one of only two beginning teachers (out of 11 participants in this study) who spoke quickly and clearly on this point, that is, who expressed an early critique. Yet she knew almost nothing about capillary rise (Viennot and Décamp 2018).

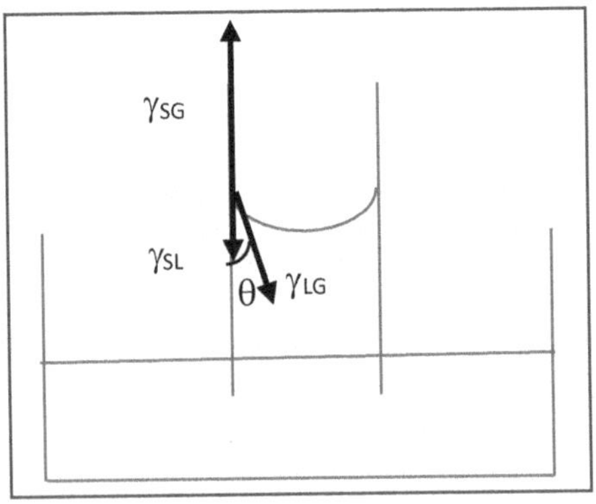

Fig. 6.3 A figure commonly used to introduce Young's formula $\gamma_{LG}\cos\theta = \gamma_{SG}-\gamma_{SL}$, where γ_{LG}, γ_{SG} and γ_{SL} are interfacial tensions (forces per unit length) for the liquid/gas, solid/gas and solid/liquid interfaces respectively, and θ is the contact angle

6.3 Conditions for Early Activation of Criticism

6.3.1 "Putting Together" Available Pieces of Information

The secret of these early critiques lies in the effort to bring together pieces of information without waiting to get all of them.

In the case of the survival blanket, the emissive power of the golden side, noted in the preamble to the interview, led to questioning of a preconceived idea, this in a person who was not familiar with the thermal transfer processes.

With regard to osmosis, thought experiments triggered early critiques in two beginning teachers in our sample. They, too, had very little knowledge of the subject matter, and were able to formulate a very pertinent critique by taking the text at face value and showing that it led to unbelievable conclusions. This does not so much imply knowledge of the text contents as a serious consideration of what is in the text.

The beginning teachers' critiques observed about the surface tension diagram for a liquid meniscus (Fig. 6.3) also illustrate an effort to reconcile basic knowledge - a given force has to be exerted on a specific object - with an unclear diagram.

These few examples illustrate the fact that the people concerned did not rely on advanced knowledge to carry out their critical analysis, but rather on a priority requirement of coherence: to try to reconcile the things that are known and ask questions without delay. Of course, the statement of this interpretation is open to discussion: what is a "thing that is known"? There is a whole range of possibilities, from familiarity with a word – such as concentration, or force - to deep understanding of a concept. To be surprised that criticism is delayed, when this is the case, is therefore to consider that the persons concerned (here beginning teachers at the end of their teacher education) had a sufficient understanding of the terms used in the text. Sometimes it turns out that the very meaning of the words used is problematic.

6.3.2 The Meaning of the Entities Mentioned in the Text

One might think that a text involving terms unknown to the reader will not be easily criticized by any reader. The first thing a reader has to do is to locate their misunderstandings of the terms used, but this is not so easy.

Metaphors can be misleading in this regard: readers think they understand. Indeed this is the goal of using a metaphor (see Chap. 3). Sometimes, it is the common usage of physics experts, especially their shortcuts, which blurs the meaning of the terms used. Students may believe, for example, that "Laplace pressure" - often mentioned in capillary problems - is a physical quantity different from "hydrostatic pressure", or "osmotic pressure", simply because an adjective (hydrostatic, osmotic) or a name (Laplace) - reminds us of the phenomenon to which a difference in values of this quantity between two points is due. Thus, in the study cited above (Viennot and Décamp 2016b) - in which future teachers were invited to discuss the theme of osmosis - some symptomatic questions about the meaning of the expression “water pressure” or of the notation p_{water} were observed.

> *Interviewer* - I asked you to define osmotic pressure?
> *Interviewee*- I can’t define it, I thought I had an explanation, but in fact (...) I can’t finish it (...).
> - What’s preventing you from reaching a conclusion?
> - Uh, the notion of pressure in water.

It turned out that when the teachers interviewed had a better understanding of this concept, they were more likely to activate their potential for criticism of the questionable aspects of osmosis explanations presented to them.

The beginning of a reading that we hope will be critical is therefore to ensure the best possible understanding of the entities that serve as a basis for explanation. This may seem obvious but, in teacher education, one should not underestimate the difficulty that certain commonly accepted terms can cause, nor the time that a reader may take before realizing their misunderstanding and asking for clarification. Our results suggest that, among those asked to give their opinion on a text, elucidation questions from the reader often announce the activation of an explicit critical attitude, but they may take some time to emerge.

6.3.3 Psycho-Cognitive Factors

Asking questions, whether about the meaning of certain terms or other aspects of a text, involves psycho-cognitive factors. This decision involves analysing one’s own understanding. It reflects a demand for understanding, a possible frustration in this regard and the expectation of coherent explanations, probably supported by a vision of what scientific knowledge is. It also implies not being paralyzed by a feeling of incompetence.

Several psycho-cognitive factors can play a blocking role, not only as regards questions of elucidation but more generally any form of critical expression. When making contact with a text, the feeling of not knowing and the desire to cling to school memories often seem to dominate, as expressed by a beginning teacher at the end of an interview on capillary rise (Viennot and Décamp 2018).

> - Well, it’s true that since we’ve been talking, I don’t feel like I understand the subject on which we’re talking, I know that I am more knowledgeable on the subject than what I can explain and that puts me in a somewhat more defensive position (…). So to say explicitly that a sentence is wrong, I feel like I don’t have the resources... that would allow me to say: no, I know that this is scientifically wrong.

The role of school habits is also undoubtedly important in blocking the questioning of a text, as in the following statement by another participant in the same study:

> -Well, in fact, it’s kind of like earlier. I’m accepting it because I’ve always done so but I never questioned the fact that it was legitimate to apply, uh, forces on an immaterial line.

Being aware of such obstacles might be a first step to progress towards a freer practice of critical analysis and reasoning in physics.

6.3.4 *Teachers and Educators: A Need for Decentering and Critical Lucidity*

The conditions that have just been outlined so that a person confronted with a text does not remain in a state of critical passivity for too long apply, to varying degrees, to experienced teachers and teacher educators.

They are strongly influenced by school habits. These rituals, which they have sometimes practiced for decades, contribute to inactivate their judgment, as we have seen with "isobaric hot air balloon" exercises.

Faced by a contestable text, some experts will not feel the need for elucidation: they already know what it is all about, in principle. They can complete or even correct a text in real time as they read it. But when they have to evaluate the effect that such a text will have on others, which is (or should be) typically a teacher's perspective, they have to activate their critical potential. They need to move away from their own sense of security, put themselves in the place of a novice by looking at the text for what it is and thus avoid expert anesthesia. This is very difficult.

It is in this perspective of teacher preparation that the next chapters are written.

References

Viennot, L., & Décamp, N. (2016a). Co-development of conceptual understanding and critical attitude: Toward a systemic analysis of the survival blanket. *European Journal of Physics, 37*(1). 015702. https://doi.org/10.1088/0143-0807/37/1/015702.

Viennot, L., & Décamp, N. (2016b). Conceptual and critical development in student teachers: First steps towards an integrated comprehension of osmosis. *International Journal of Science Education, 38*(14), 2197–2219. https://doi.org/10.1080/09500693.2016.1230793.

Viennot, L., & Décamp, N. (2018). The transition towards critique: Discussing capillary Ascension with beginning teachers. *European Journal of Physics, 39*, 045704.

Chapter 7
An Education in Critical Analysis

This is undoubtedly the most ambitious chapter title of the book, though realism requires us to be modest. However, it is not forbidden to have challenging objectives. As regards education, it is important to broaden the scope considerably. Up to now, examples of texts requiring critical work have concerned statements and/or images presenting or suggesting flaws from the point of view of expert physics. It is necessary to pay particular attention to the possible consequences of such texts for novices. But this should not make us forget the potential benefit of a critical analysis of texts that *do not have* such weaknesses. Any document with an informative and explanatory purpose is worth questioning if we want to learn from it. It is therefore important that education in critical analysis for oneself or others makes it possible to make the most of any text. This ability, which is often advocated for responsible citizenship education, is essential in the field of scientific instruction. The vision of the solitary researcher who conducts his research by managing reasoning, experiments and conclusions alone does not hold for a second. Knowing what others are doing, or more specifically knowing it critically, is absolutely essential.

7.1 Maximising the Benefit of a Text

The categories we have detailed above as grounds for rejecting (Chap. 2) or questioning (Chap. 3) a text are useful because they can help us to evaluate an argument. Their aim is to arm those who want to learn to avoid being misled and to guide the pedagogical choices of those who want to teach. But even with a text that does not contain obvious errors, maximizing its benefits presupposes a thorough critical analysis. It is in this spirit that we return here to certain aspects pointed out above, knowing that they could all be treated in the same way.

L. Viennot, N. Décamp, *Developing Critical Thinking in Physics*, Contributions from Science Education Research 7,
https://doi.org/10.1007/978-3-030-43773-2_7

7.1.1 The Meaning of the Terms Involved in the Explanation

It is essential for learning or teaching to pay attention to the meaning of the terms and expressions used in a text, even if these are not particularly ambiguous or metaphorical. For example, our students have a vested interest in understanding that when we talk about a *uniform* field we mean that it is the same everywhere at a given time and when we talk about a *constant* field we mean that at a given point it does not change with time. They also have an interest in understanding that the electromotive force is not a force and is measured in volts or that the expression *material point* deserves to be carefully interpreted. Although such a phrase is so classical that it is a standard of the last year of secondary education in France (Bulletin de l'Education Nationale 2011, p. 83–97), it nevertheless generates misunderstandings, as we have seen in connection with surface tension forces (Chap. 4, see also Appendix G). One could cite many examples of very common designations in physics that, in fact, deserve a very precise explanation.

7.1.2 Incompleteness in the Form of Implicitness

It is important to uncover implicitness. Indeed, clarifying the meaning of a term or expression that is overly implicit often diminishes the logical incompleteness of a text. The following text (Jacquier and Vannimenus 2005), for example, has no errors:

> The Nobel Prize was awarded in 1997 to Claude Cohen Tanoudji, Steven Chu and Bill Phillips for demonstrating the principle of cooling atoms by laser at extremely low temperatures, of the order of one thousandth of a degree above absolute zero. [...] Cold atoms are now used in atomic clocks that measure time with extraordinary precision[...](p. 136).

For a physicist, it goes without saying that being cold means, in the case of atoms, having a low velocity in the reference frame of study. But a layperson will have a better chance of understanding something if they are informed of the implicit link here: The interest of cold atoms is that they are slow, a fact that prevents frequency dispersion of their radiation by Doppler effect. For a demanding reader, it is therefore beneficial to ask for a logical link between the coldness of atoms and the precision of atomic clocks. Thus, a well-understood critical analysis allows progress in understanding.

This is particularly the case when a text mentions a constant or a coefficient. It has been shown (Viennot 1982) that the statement "the speed of light in vacuum is a constant" is not commonly understood in the sense that it does not mention what this speed is independent of. Adding that this speed does not depend on the Galilean reference frame where it is measured gives this statement a scope that is all the more interesting as something surprising is made explicit.

Another example in this register concerns friction. An exercise text notes μ_d the dynamic coefficient of friction between an object and the surface with which it is in contact, which allows us to write the value of the force arising from the friction interaction as $F = \mu_d N$ where N is the normal component of the contact interaction.

When N is equal to the normal component of the weight, and if all the other forces involved are proportional to mass (m), then, by simplification, the mass disappears from the fundamental equation of dynamics $\vec{F} = m\vec{a}$. As a result, acceleration ($\vec{a}$) and consequently the movement of the object concerned does not depend on its mass. But a critical look at this text could lead us to wonder: what about μ_d? Is this coefficient independent of mass? of speed? of the contact area? It is interesting to note that a commonly used model answers all these questions positively, but that this is not obvious. This discussion enriches the initial information, it leads to assessing the advantages of the simplified model while pointing out its possible limitations. We can also understand that if downhill skiing champions are generally quite massive, it is because another force acts on them, a force that is *not* proportional to the mass, i.e. the viscous friction force exerted by the air (Viennot 2014, p. 28).

7.1.3 *Generalization: Not Necessarily Wrong*

As we have just observed, the in-depth analysis of a text can lead to an interesting generalization: when all the forces exerted on an object of mass m are proportional to this mass (m), the movement of the object in question does not depend on it. Thus, when we examine the extent to which a conclusion can be generalized, we sometimes end up with an interesting surprise. Do we remember all these charged particles (q) that circulate, exercise after exercise, in regions where the magnetic flux density $\vec{B}$ is said to be uniform? They accustom us to the fact that their speed ($\vec{v}$) is constant over time. The reason for this is that the force exerted on them, in the absence of an electric field, is given by the relationship $\vec{F} = q\vec{v} \times \vec{B}$. The cross product in it ensures that this force is perpendicular to the trajectory. And so, surprise, in this model the affirmation of an unaltered speed can be generalized to any magnetic field, however non-uniform it may be. We only become aware of this interesting result by questioning the starting hypothesis of the exercise, i.e. the uniformity of the field: what changes if we change the hypothesis? This is a critical analysis, but it does not call into question the validity of the original text. It's about how far you can exploit it, how to maximize its benefits.

7.1.4 *From Numerical to Functional: A Big Step Towards Better Understanding*

In fact, progress towards a better understanding often depends on a change in the reading of formulas and expressions of physics. Seeing constants as numbers, or relationships as equalities between numbers, allows some numerical values to be calculated from others. But considering these elements as expressing dependencies or non-dependencies goes much further. As we have just seen about solid friction, it will be possible to realise that some coefficients remain appropriate over a wide range of phenomena. It will be possible to specify limiting cases, for example for an extreme value of a variable, the others being fixed, and to examine their plausibility. It can even be argued that without this approach based on functional understanding, there is a risk of having little or no understanding of a scientific text.

7.2 Critical Analysis: An Activity that Is Always Productive

If we want to instill the virus of critical analysis in any audience, it is probably useful not to be in a purely defensive posture against error, imposture or negligence. The power of rigorous and imaginative critical analysis – inventive of thought experiments – is to control but also to extend the informative contribution of the text in question, for better comprehension.

7.2.1 School or University Environment and Classical Text

It is in this sense that, in a school or university setting, the activity of analyzing course texts or solved problems is highly recommended as an activity in its own right. For example, you can give both an exercise and the corresponding expected answers to students and ask them a series of questions (examples in Table 7.1) about these documents (Viennot 2014, Chap. 5). Freed of the concern for calculation, they will have to concentrate on the meaning and scope of the text.

This critical reading work is particularly essential in a popular learning format: the *Flipped classroom* (Bergman and Sams 2014). Here students are required to consult certain documents (including online courses) before meeting with the teacher or trainer. This presupposes, on the part of the target audience, an autonomous active attitude from the first contact with these resources: it is a question of appropriating them. Thus, if someone wants to learn, he must ask him/herself the questions suggested above: meaning of the entities present in the statements, location of ambiguities conveyed by the text, an image or an analogy, internal or external coherence test, detection of missing (or apparently missing) logical links, analysis of functional dependencies and proposed or potential generalizations.

Of course, we must also accept the frustration of not knowing how to respond immediately: that is the role of the teacher who will advance the understanding of their students through activities that really involve them. It can be argued that this is a wishful thinking. Yet, it has been observed that appropriate education and long-term promotion of critical attitude can increase the chances of success (Bergman and Sams 2014).

Table 7.1 Topics or suggestions for questions about a solved problem or a passage in a physics textbook

The meaning of symbols or verbal expressions used in the text being analysed
The influence of 'coding' in designating symbols and in writing down algebraic relationships in the text: this section relates to what is often described as the 'sign convention'. One might wonder how the way a calculation is written alters when the orientation of an axis is changed (e.g. 'z axis')
The assumptions made in the text: here the very foundations of the argument are examined. This section deals with often difficult questions, for example:
Where does this assumption come in?
On what assumption is such and such an assertion based?
What must be altered if this assumption is changed?
Analyse the orders of magnitude which justify this hypothesis.
Calculation:
Is such and such an equality proven in the text, or simply 'parachuted in', e.g. '... it is well known that...'?
Where are the simple intermediate steps of the calculation?
The result. What is involved here is the significance of the conclusion:
State the result in sentences
Check the result: consistency and limiting cases.
How does such and such a variable vary when another increases, the others remaining unchanged?
Obtain an order of magnitude for a given variable in a given circumstance.
Extension: slightly extend the calculation given in the text to draw out extra information.

What is valid in a school or university setting is also valid in a situation of personal learning or consultation of popular texts, in a search for comprehension.

7.2.2 *School Setting and Text with Intermediate Status*

In a school environment, critical analysis may also be conducted with a less formal text, for instance one using a motivating context. Such is the case of the document "ETs" (Fig. 3.16) already commented on in Chap. 3 (p. 52). An investigation by Feller (Feller et al. 2009) gives access to the effects of a discussion in class about this document, supplemented by the following text:

> The scene takes place in the year 2000. One may wonder whether the alien can answer the question. In fact, if the Alien knew when mankind domesticated fire, he could deduce that the Earth is 300,000 light years from their planet, that is, about 31.10^{18} km. This shows that the propagation of light is not instantaneous.

Table 7.2 illustrates a particularly significant moment.

Table 7.2 Excerpt from a class discussion about Document "ETs" (Feller, Colin and Viennot 2009, p. 80; Viennot 2014, p. 153)

Transcript	Our comments
Teacher: **– There are only two inhabitants on the ET's planet.** What do you think of this response?	*The teacher points to an implausible element*
A: *– It's true, but it's a drawing, we are not going to represent all the people on Mars.*	*This student considers it to be anecdotal*
T – All right. The question is to know if it hinders the comprehension of the main message.	*The teacher raises the question as to whether this is a potential obstacle to understanding?*
All – *Well no, it doesn't matter!*	*It is not considered by the students to be consequential for comprehension*
(…)	*(Same type of dialogue about two other students' responses: the teacher lets the students talk freely.)*
	The teacher submits a more questionable proposal for discussion
T – Yes, all right. Wait, I'll read (a student's sentence previously collected) : **It is not possible to see these details (the Earthlings are hairy and they are trying to make a fire) with a telescope, especially if we're 300,000 light-years away.** So, what do you think of this? Can this blur the main message?	
B – *Well no! Scale isn't the point for the author. A big or small telescope, it doesn't make any difference*	*Spontaneously, the student turns to the problem of scales*
T – All right.	
C – *Well, look at response (6). He says that the scale is incorrect. But it is in no way the purpose of the author to have appropriate scales.*	*These students (B, C) seem to consider the issue of scales to be unimportant*
D – *Yes, we aren't going to draw one planet and then the other one thousands of light-years away; there isn't enough room.*	
B – *In fact, this way of showing us things helps us understand.*	
T – All right, so, I ask you: what do you think of the student who said: we get the impression that the ETs are close to the Earth?	*The student (D) takes into account the illustrator's point of view. But the potential obstacle associated with the inappropriate scaling is not identified*
D – *Well yes, it's the only planet that we can see with he naked eye.*	
T – Well then, if we are so close to the Earth, is the light going to take 300,000 years before reaching us?	*The teacher tries to orientate the discussion towards the representation of the Earth*
B – *Well no, it will take much less time.*	*Gradually, the teacher stresses the inconsistency linked to the size of the Earth*
T – Then let us pose the question the another way: if we had represented the Earth at a distance of 300,000 light-years, what would we have seen?	
All – *Well, nothing!*	
E – *A point.*	
T – Then why did the author represent the Earth like that?	
F – *Well it's so we'll understand that it's the Earth.*	*The student (F) underlines the reason why the drawing is like that: in order to facilitate the identification of the Earth in the image*
T – All right, so if we draw the Earth as a point, can we understand the main message all the same?	*This reason is soon challenged.*
G – *You bet we can, it is what they say that enables us to understand: I see Earthlings …*	
T – All right. So, was it necessary to represent the Earth like that?	
All: *– No!*	
T – You told me first of all: the scale isn't a serious problem. What do you think now?	
H – *Well yes, there, it does matter*	

(In bold: some students' sentences previously collected)

This discussion bears witness to several episodes of student awareness, including about the constraints of the graphic artist: "The Earth had to be recognized". Moreover, in terms of comprehension, the students were able to put into perspective the consequences of the choices made by the cartoonist. Reaching this level of critical maturity at grade ten level is therefore possible, at moderate time cost. Despite the difficulty of this type of activity, the participants at the end of their third and last session (this time done on their own) were very positive.

> - I found this work very instructive; it enabled me to study a document in depth and to write a critique of a popular text. I think that a document of this type should contain simple but clear explanations. When some notions are simplified too much, they sometimes end up being barely understandable. (…) Doing this work, I also realised that the goal of physics was to extract relevant pieces of information from a document, to select the main ones, and not to simply make calculations or experiments."

> - This work helped me to develop my critical sense. Now, I better understand why it is necessary to consult several sources to be sure about the pieces of information that we talked about. So it is interesting to analyse a document, and to know exactly what we do or do not understand.

> -This work convinced me that it is necessary to take time to read an article, and not just glance through it, because it may be very interesting. We should not give up because of something we do not understand. It also made me understand that I could be interested in a subject which, at first sight, would not have attracted me if I had not been obliged to work on it (...) After all, this work made me much more aware of things I would not have thought of otherwise.

Yet, what was observed at the beginning of this sequence was a typical syndrome of *delayed critique*: responding to a written questionnaire, only 9 students out of 94 had pointed to the fact that the image of the Earth was too big. Additionally, none of them had signaled that this might have hampered the comprehension of the document.

7.3 The Education of Future Teachers: Awareness and Reluctance

In order to provide the kind of education just mentioned, we need well-informed teachers. We can note that the episode we have just described is based on an official document of the Ministry of National Education in France (Centre National de Documentation Pédagogique 2000) presented without any particular warning. This reflects a remarkable absence of critique, given the number of experts in the group promoting this text. There are therefore good reasons for explicitly including critical analysis in teacher education. There is no question here of offering a model for such an educational intervention, but rather of proposing a few pieces of information that may be useful to those who would venture into this relatively unexplored field. These elements may be used in long (ideally) as well as in very limited sessions (2 or 3 days, as is often the case with initial or in-service teacher education programmes).

Firstly, given the psycho-cognitive aspects involved in critical expression, it seems desirable that participants in such sessions should be given the opportunity to experience for themselves the situation of a potentially critical person faced with a text. Ideally, if we can propose a text to each participant and ask for their opinion on it, their personal awareness can be raised: "Why didn't I react earlier? I could see that something was wrong, but I didn't say anything"; "I had seen this question a long time ago and I had forgotten everything, so I didn't dare to talk." These are all statements that may emerge in such circumstances. The management of training sessions can take other forms, such as mini working groups with exchange of roles between those who propose a text to be criticized and those who undertake to analyze it.

Secondly, an activity involving the characterization of questionable texts is *a priori* useful to sharpen participants' subsequent vigilance. For this purpose, it is necessary to have both a bank of typical texts and an analysis grid. The examples compiled over the course of this book can contribute to this (Appendix J), in association with our dual analysis grid: that of manifest flaws (Chap. 2) and risk factors (Chap. 3). However, this does not mean that classification should be turned into a formal exercise. As with any classification, the point is not necessarily to lead to clear-cut and indisputable decisions. Rather, it is a matter of enriching the repertoire of questions that can be asked concerning a text, in order to better understand its scope and possibly to challenge it.

Finally, it is very important to discuss with the participants how their new critical experience might influence their teaching practice. Table 7.3 reproduces a dialogue at the end of the first day in a two-day workshop with beginning teachers at the end of their programme (the texts used in this workshop are all analysed in this book).

Obviously, awareness has increased and the need for vigilance is expressed. These are grounds for optimism. A comment indicating a good understanding of a topic dealt with in the course is also encouraging:

> - There is one thing (...) to be wary of, and that is any explanation that comes close to a story. It happened to me yesterday, (...) because I had just had the previous warning on this issue, I started an explanation like a story and then I said to myself: stop!

However, one should not underestimate the obstacles that participants themselves expressed in relation to their classroom practice.

7.4 Towards Classroom Implementation?

7.4.1 A Feeling of Difficulty

It is quite clear that the activity of text criticism appeared difficult to this sample of beginning teachers.

Table 7.3 A discussion between a teacher and participants after a day in a workshop on critical analysis

Educator- You say that you have really become aware of the need to allow yourself to criticize even if you have not understood everything?
Participant 1-Ah yes!
(…)
Participant 2- Yes, yes, using a text, you really have to take your time to analyze it (…), make cuts and additions
Educator- Or present it as it is and, with the students, see what you need to add or remove. Is that too ambitious? Is that why you're shaking your head?
Participant 3- Above all, we must have them all identified, all identified (*these contestable arguments*). I would be happy to ask them (*students to criticize*) but if I haven't seen what should be criticized …
Participant 1- Is it that bad? Actually, you made progress today, [the next time] you will face the situation and the more you are confronted with a critical situation live, the better you will be able to manage it, react ever better

- We both came out saying that we were unable to analyse the texts as you did, there are many things we hadn't seen.
- We don't necessarily realize that there can be problems.
- I'm in seriously questioning my own abilities, because I feel like I'm not capable of doing this on my own.
- It's just that I can't manage to do this on my own. I can't stand back when I'm planning my class, it comes afterwards.

The participants did not stop at this observation but clearly analysed some of the reasons for their acritical attitude.

7.4.2 *Demonstration As It Should Be*

One of the reasons for accepting some of the proposed texts without proper analysis is that they can take on the appearance of a perfectly suitable demonstration. Thus, with regard to the limewater test, the classical sequencing of logical connectors (I know that, therefore) is likely to give confidence:

Participant – It resonates in my head.
Educator – It sounds like a demonstration, it's meant to be a demonstration.
Participant – It's meant.

This anesthetic effect is accentuated when the conclusion is correct.

7.4.3 It Works, So It's All Good

The *"aggravating circumstance"* referred to in Chap. 3 is clearly mentioned. When an explanation seems to lead to a correct result, when *"it works"*, any criticism seems inappropriate.

- At some point, you're so desperate to show this thing that you find the thing that works.

Quite close to this register, one participant's commentary on the limewater test is the sober expression of a resource optimization strategy.

- There's only one door, there's only one key, I have a key, so I open the door.

7.4.4 Classroom Management

The participants also justified their obedient acceptance of certain explanations for reasons of economy in their class management.

A first idea expressed is that, as long as students do not perceive the flaw in a line of reasoning, there is no disadvantage in using it. This can be supplemented by the simple desire to minimise the teaching effort, as long as insecurity remains at an acceptable level:

- I think that as long as I'm not confronted with the problem, as long as the students don't put the problem in front of my nose, that I'm not in a delicate situation, well, (…) that's why I have trouble doing the work upstream (...)..

But it can also happen that a participant proposes not to follow up on the critical remark made by an insightful student, on the grounds that it is necessary to "move on":

- I have a tendency, when I want to move forward, I rather tend to criticize this critical spirit a little by saying good, you saw the pitfall, but now we have to move on, today's model is more restricted and you will see that later, whereas yes, he felt, it would be interesting to discuss this.

There is also the fear of "confusing" or "frustrating them":

- Yeah, but if we tell him a lot of things and then in the end we tell him well... finally, that's not good enough. …

And some propose the idea of compromise:

- There are some things we find in books, I call them lies for children, we know that it's simplified, we know that it's not completely accurate, but we can't explain at their level why it's not accurate, it's just, it's something that's going to work in their world and then when their world is bigger, then we can provide them with a better (...).

- Here we come to the notion of a model and if you want to be honest with the children, you have to tell them that it is a model, not exactly the most precise model that exists at the moment.

It is to be noted that the limits envisaged in these last remarks are expressed in terms of greater or lesser simplification, in relation to the concept of model. As is often the case, the idea of coherence is less emphasized than that of precision.

Among the participants' comments, we also find some regrets about the education they themselves received.

7.4.5 Critical Attitude and Education: The Views of Future Teachers

Returning to their own career paths, the participants note the links between the development of a critical attitude and their own education.

- Education is what causes it (*critical passivity*), I think it's one of the causes, because we don't allow ourselves to be critical.

- There are a lot of things that we hadn't seen and I think it doesn't just come from the lack of practice, maybe there are some things that are internalized by teachers, the important thing is to learn to do it consciously. (...). We haven't been educated to do that.

One can therefore count on reactions of these types if one embarks on an education of teachers in critical analysis. Achieving this level of awareness seems to be an attainable goal in a relatively short time. But to promote a real change in classroom practices, the road is certainly longer.

References

Bergman, J., & Sams, A. (2014). *The flipped classroom Repentigny*. Québec: Reynald Goulet.

Bulletin Officiel de l'Education Nationale. (2011). special issue n°8, October 13th 2011, NOR : MENE1119475A.

Centre National de Documentation Pédagogique (France). (2000). *Document d'accompagnement du programme de Seconde Générale* (grade 10).

Feller, I., Colin, P., & Viennot. (2009). Critical analysis of popularisation documents in the physics classroom. An action-research in grade 10. *Problems of education in the 21st century, 17*, 72–96 http://journals.indexcopernicus.com/abstract.php?icid=900326

Viennot, L. (1982). L'implicite en physique: les étudiants et les constantes. *European Journal of Physics, 3*, 174–180.

Viennot, L. (2014). *Thinking in physics, the pleasure of reasoning and understanding*. Dordrecht: Springer.

Chapter 8
Critique: A Prelude to Deeper Comprehension

Why is it so hard to teach critical thinking (Willingham 2007)? This question is not very encouraging. There is a lot of talk about developing "critical thinking" in science education, but far too little is being done: it is future teachers who say so, and they are among the good students.

Willingham's question is a useful caution, but can we seriously decide to neglect critical thinking when we are a teacher or teacher educator?

Undoubtedly, an important prerequisite for the development of critical analysis is to think of scientific learning as a journey (or a hike) for which both learner and teacher are co-responsible. So learning is not (only) "following", it is not (only) being able to repeat, like showing pictures after our return. It means looking carefully at where you put your feet, and how you use your eyes and ears to allow the next step forward. In this attitude, critical analysis ensures the steps; it makes it possible to know what to count on, to get the best out of the information consulted. This is not only true for the self-taught, because fortunately there are experts and it would be foolish not to take advantage of them. To stay in a metaphorical register, when a mountaineer wants to check a hold, they pull or push on it while maintaining their other supports. That's what you do when you question a text. The question is not to denigrate it but to see if, and to what extent, "it holds". Hopefully, there will be good support for extending the conclusion of the text, or asking a good question at some stage. So our aim is not primarily a trial.

The abundance of examples provided in this book may surprise, given their sometimes very academic references. Moreover, the more we are forewarned of what can happen, the more discoveries multiply. Texts that lend themselves to rebuttal or doubt on a rational basis are legion, and we soon realize that we produce some of them ourselves on occasion (the reader will be able to practice on this book!). That's the way it is. The human mind is able to trace carefully crafted argumentation paths, very well controlled by the scientific community. But daily life, typical teaching, ordinary popularization, rapid expression, audiences of different levels do not allow us to remain easily on these summits of rigour. Common reasoning in physics

L. Viennot, N. Décamp, *Developing Critical Thinking in Physics*,
Contributions from Science Education Research 7,
https://doi.org/10.1007/978-3-030-43773-2_8

is widely shared. It is the place of resonance between those who explain and their targets, by a conscious or unconscious mechanism. In particular, these *echo explanations* impose a causal linear structure to varying degrees, making binary links between variables, which is not appropriate for systems analysis. We must recognize that we generally come to understand each other on the basis of a somewhat imprecise expression. But it is worth examining the cases where this does not work, those where the silent failure of understanding goes hand in hand with the collapse of coherence.

A crucial issue is the awareness that the person receiving an explanation has, or does not have, of their lack of understanding. Such a perception is a privileged gateway to critical analysis. On the other hand, we have observed a phenomenon that we have termed "anesthesia of judgment" |(as discussed in Chap. 5), which affects those who are familiar with a given subject, when faced with very incomplete or even inconsistent texts: these "experts" confuse the fact that they understand the subject with that of understanding the text. Focusing on themselves rather than on the explanation proposed, they correct in real time what they see and are satisfied with the reminders that this reading gives them. They are therefore not in a position to anticipate the effect of the text on others. To a greater or lesser extent, we are all likely to find ourselves in this situation at some time or another. This lack of critical analysis – or *expert anesthesia* – can be seen as one of the factors responsible for the survival of apparently indestructible teaching rituals, isobaric balloons and other materialized rays, which continue to appear in texts without being questioned.

The absence of criticism is often due to a completely different phenomenon: *delayed critique*, observed during in-depth interviews with prospective teachers faced with a text. What dominates at the beginning of the discussion is not awareness of misunderstanding, but a search for fading school memories. It takes time to reflect, and an input of information from an expert, to trigger a truly critical attitude towards the text under discussion. We can speak of a threshold of comprehension necessary to liberate criticism, even where this is not logically necessary. In fact, to detect the flaws in a text, it can be enough to pay serious attention to the text, such as the double assertion that radiocarbon (^{14}C) disintegrates and that its concentration (relative to ^{12}C) remains constant in the atmosphere. This leads to a two-fold conclusion.

First, without conceptual mastery, it is less likely that we will see an expression of critical analysis. People who want to learn or be informed focus their efforts on comprehension of the topic addressed rather than criticism of the text, even if these two activities should be mutually supportive. With the intellectual insecurity that comes from their sparse memories, they prefer to trust the text and try to understand what they read rather than question it. Here is a good argument against a vision of critical thinking development as a skill in itself, not supported by a conceptual construct. It also supports the idea that "Critical thinking is not a set of skills that can be deployed at any time in any context" (Willingham 2007, p. 10): a person is in a different situation with respect to his or her threshold of comprehension depending on the subject concerned. This argues against the use of the expression "critical thinking" which may suggest an intrinsic quality of a given individual. We prefer to speak of a critical attitude as the activation of a potential that cognitive psychologists assure us is already present in very young children (Willingham 2007).

On the other hand, we may wish to avoid excessively inhibiting criticism. When taking a text seriously, without any real problem of meaning of the terms used, is enough to detect an inconsistency or a logical incompleteness, why wait? When presented with an explanation that is in principle intended for them, our students, their teachers, and ultimately everyone would benefit from an honest inventory of what is explained and what is not. The intellectual requirement must be combined with rigorous analysis and that is what this book would like to encourage. But is it really "so hard to teach critical thinking"?

There are a number of reasons for optimism. First of all, it is true that people may well show an early critical attitude to texts on subjects of which they are far from being specialists, without being constrained by their own ignorance. It is rare, but it does at least show that this goal is attainable. Second, almost everyone interviewed individually in the investigations discussed in Chap. 7 showed great satisfaction at the end of the interview even though they had to question their own reluctance to be critical. The same phenomenon occurred in educational sessions with groups. Prior to this final satisfaction, surprise was more widely expressed – "This is the first time I asked myself this question" – often accompanied by a critique of their previous education. Often coming after a long period of perplexity, this triggering of a critical and lucid attitude is a significant event: "I am in violent questioning", said one participant. Another testified that during his teacher education course he recognized a risk factor identified during on the previous day of his course:" I started an explanation like a story and then I said to myself: stop!"

At least two conditions seem necessary for an education involving such events to have lasting effects. One, already mentioned, is that a minimal conceptual framework has been developed for the topics covered. The aim is to provide the person in the course of their education with references regarding the meaning of the terms used and practice in logical sequences of reasoning. The other condition is psycho-cognitive. Everyone should believe that they are entitled to expect an explanation that makes sense when the author claims to provide it. The reader's scattered knowledge and faulty memories do not justify the passive acceptance of statements that they have, in fact, all the means to challenge. Having measured in our surveys how some future teachers delayed their criticism, going back several times before expressing it, we are tempted to say that there is a trace of major psycho-cognitive blockages. From this perspective, teaching critical analysis would not only involve scrolling through a series of examples in front of trainees by soliciting their critical opinions. The idea that it is necessary to train our prefrontal cortex to take control of thought is interesting (Houdé 2014), but still not very operational. A crucial question is to identify cases where this control is particularly necessary. Sinking into a permanent critical paranoia would dangerously encumber our neuronal machinery. Hence the value of having a grid of types of flaws and risk factors that allows us to direct our critical attention selectively. It is at least as important to convince everyone that they can use their critical potential, that they have the means to do so and that they are allowed to do so.

This effort is more likely to be successful if critical analysis avoids a purely negative register. Indeed it is essential to see critical analysis as a springboard for action, and this remark goes far beyond the realm of physics. Seeing what is wrong, seeing what is missing, is a rational basis for moving forward, by imposing the same exigency on oneself. In physics as elsewhere, this way of working towards a better understanding is not only valid for texts with inconsistencies or major gaps. In fact, critical analysis is the best way to make the most of any text with an explanatory purpose. The aim is to extract the information provided as precisely as possible, to assess the value of the logical connectors, to test the generalizations it suggests, in short to evaluate what this text can really be used for. As the basis of this work, connections play a major role. The beginning and end of a text, formal approach and natural language, several pieces of information from various sources: all of this must be integrated without contradiction. Sometimes, when the benefits are there, we find intellectual satisfaction that is worth more, in terms of motivation to learn, than staging effects and splendid images.

At the end of this book, we wish to reaffirm the importance of developing intellectual life on a twofold level: conceptual and critical. In philosophical matters, this goes without saying. The reader is therefore invited to generalize this approach. Beyond the knowledge that each individual accumulates, or rather, based on this foundation, critical analysis is an essential component of thought. The modest but free exercise of this right constitutes a magnificent educational objective, which deserves a real commitment from institutions and individuals.

References

Houdé, O. (2014). *Le raisonnement*. Paris: PUF.

Willingham, D. T. (2007). Critical thinking: Why is it so hard to teach? *American Educator, 31*, 8–19.

Appendices

Appendix A. An Epistemological Position

Our approach to critical thinking in physics is based on an epistemological position: any group seeking to develop new scientific knowledge aims to produce statements and submit them to critique, either internally or by opening them up for wider discussion. Thus, physics is a field, like others, in which critical analysis is essential. However, some factors seem to us to be particularly favourable for learning critical analysis through physics.

The first is related to the high degree of mathematical formalisation that is the rule in physics. Most statements produced in physics are in the form of mathematical relationships, which is an aid (as long as you have mastered mathematical language) to detecting internal contradictions or becoming more aware of the role of a particular variable in a piece of reasoning. Indeed, as Hulin (1992) argues, "natural language alone (...) is ill-suited to express the relationships between physical quantities, let alone to develop an argument based on these relationships".

A second favourable factor is the existence of well-defined and measurable quantities in physics. Measurable quantities do of course exist in other fields, but in physics this is the rule and this omnipresence of numerical values makes it possible to test predictions quantitatively. As Ogborn (1999) argues, science is constrained by a continuous effort "to make tight connections between transactional imagination and intransigent reality" which means "taking seriously and systematically developing the simple and obvious thought that although we can think whatever we like, we cannot do whatever we like" (p.17). While it is relatively easy to detect a discrepancy between an expected result and an observed result, it can become more complicated to determine how to explain it. Such a discrepancy may result from the fundamental physical laws at play, from the model of the physical situation, or from the measurements themselves. Fundamental physical laws are rarely challenged directly. In fact, physics is a very structured science aiming at a coherent and

L. Viennot, N. Décamp, *Developing Critical Thinking in Physics*, Contributions from Science Education Research 7,
https://doi.org/10.1007/978-3-030-43773-2

parsimonious description of the world, a few laws accounting for a large set of phenomena in a specified range of validity: "(…), the value of explanations resides in their enabling us to unify and to organize knowledge. (…). Scientific knowledge is thus unified to the extent that we can derive the largest number of facts from the smallest number of assumptions" (Norris et al. 1989, See also Kitcher 1981; Ogborn 1997; Jenkins 2007; Thagard 2008). This perspective of unification provides essential guidance for deciding what to conclude in the event of confirming evidence or of discrepancy between a prediction and the outcome of an experiment. For instance, Papadouris et al. (2018) have recently taken up Duhem's position (1908): "Not withstanding the critically important role of observational data in scrutinizing explanations, they do not suffice to yield conclusive validity judgments. Appreciating this subtlety is important in that it allows putting empirical evidence in an epistemically informed perspective that avoids overestimating its role" (p. 224). The case of the discovery of Neptune in 1846 is often discussed in this regard (Holton and Brush 2001). A discrepancy between prediction and observation had occurred regarding the trajectory of Uranus. In this case, given the great unifying power of Newton's law of universal gravitation, it was finally decided to seek a new planet rather than reject the theory. In some cases, however, it is the physical law that must be reviewed and not the model (which includes here the number of planets to be taken into account). This then leads to the restructuring of entire parts of physical theory to ensure the coherence of the whole (see the "scientific revolutions" defined by Kuhn 1962). Such restructuring occurred, for example, during the transition from Newtonian theory to general relativity, which now explains the advance of Mercury's perihelion.

In this book, it is this great coherence of physics that we wish to highlight, and the constraints it brings to reasoning, whether in terms of explanations or predictions. In such a framework, requirements are basic: a minimum critical faculty means being able to detect self-contradictory statements or statements that contradict basic laws of physics, and also to identify explanations that are incomplete from a logical standpoint. These contradictions or incompletenesses may be detectable with more or less simple reasoning. In our case, we envisage situations where questionable statements can be identified from a very simple argument, "simple" meaning what is reasonably expectable from novice teachers and even younger students. This emphasis on coherence does not mean that this concern is a sufficient condition for the progress of science. As Ogborn (1999) wrote "Nothing about the process of science guarantees success" (p. 6). But at least can we say that identifying a defect in a particular argument – in terms of coherence or logical incompleteness – is a crucial capacity in the practice of science in general and in physics in particular. This capacity does not cover the whole range of meanings of the expression "critical thinking" (Jiménez-Aleixandre and Puig 2009) but it is an essential asset, a common (minimum) condition for a productive intellectual life whatever the domain.

With regard to physics, one risk deserves to be highlighted. As Ogborn (1999) notes, "There are (…), some areas of knowledge where we may properly speak of practical certainty, that is, of knowledge on which it is appropriate to rely without hesitation until further notice. Doubt vanishes because no serious alternatives remain unexamined and uneliminated. It is probably best to think of these areas of secure knowledge as partial islands in an ocean of unknowing or of partial and insecure knowledge, though the value of their achieved existence is such that we are always liable to overestimate their magnitude" (p. 6). We must bear in mind that whatever explanation we accept as valid must be considered as an intellectual object open to improvement (Ogborn 1997; Papadouris et al. 2018), if not to radical change (Kuhn 1962).

Another risk concerns the idea of simplification. It is clear that explanations in physics are based on models of the material world that leave out some of its complexity. Hence the idea that simplification is both inevitable and fruitful, and that it is therefore legitimate as a first step in the exploration of physical phenomena. But we would stress that certain types of consequences cannot be avoided in a given model, which limits the scope of what can be successfully considered as a preliminary simplification. The emblematic case of the "isobaric hot air balloon" discussed in this book is enlightening in this respect, because the envisaged simplification ruins any hope of a coherent explanation by eliminating a crucial variable: the pressure gradient.

These constraints and risks facing science make critical thinking an essential condition for the development of scientific explanations and the learning of science. As Henderson et al. (2016) pointed out: *(…)* "critique is essential to evaluating which idea is more fruitful, plausible, or simply parsimonious" (p. 1683). We share their view that "Constructing knowledge (…), is not so much a product of one conception being replaced by another (e.g., Posner et al. 1982), but rather a process of weighing alternates and evaluating the balance of probabilities between two competing beliefs – in short – 'a dialectic between construction and critique' (Ford 2008)" (p. 1676). We acknowledge that, with regard to this evaluation process, the considerations we propose here concern only very preliminary steps, i.e. a check on internal consistency, compatibility with widely accepted laws and the logical coherence of explanatory texts.

References of Appendix A

Duhem, P. (2003). *Sauver les apparences*. Paris: Vrin (original edition: 1908).

Ford, M. J. (2008). Disciplinary authority and accountability in scientific practice and learning. *Science Education, 92*(3), 404–423.

Henderson, J. B., MacPherson, A., Osborne, J., & Wild, A. (2015). Beyond construction: Five arguments for the role and value of critique in learning science. *International Journal of Science Education, 37*(10), 1668–1697. https://doi.org/10.1080/09500693.2015.1043598.

Holton, G., & Brush, G. S. (2001). *Physics, the human adventure: From Copernicus to Einstein and beyond* (3rd ed.). New Brunswick: Rutgers University Press.

Hulin, M. (1992). *Le mirage et la nécessité: pour une redéfinition de la formation scientifique de base*. Paris: Presse de l'Ecole Normale Supérieure et du Palais de la Découverte.
Jenkins, E. W. (2007). School science: A questionable construct? *Journal of Curriculum Studies, 39*, 265–282. doi:https://doi.org/10.1080/00220270701245295.
Jiménez-Aleixandre, M. P., & Puig, B. (2009). Argumentation, evidence evaluation and critical thinking. In B. J. Fraser, K. Tobin, & C. McRobbie (Eds.), *Second international handbook of science education* (pp. 1001–1015). Dordrecht: Springer. doi:https://doi.org/10.1007/978-1-4020-9041-7.
Kitcher, P. (1981). Explanatory unification. *Philosophy of Science, 48*(4), 507–531.
Kuhn, T. S. (1962). *The structure of scientific revolutions*. Chicago: University of Chicago press.
Norris, S. P., Guilbert, S. M., Smith, M. L., Hakimelahi, S., &, Phillips, L. M. (2005) A theoretical framework for narrative explanation in science, Wiley Periodicals, Inc. *Science Education, 89*, 535–563.
Ogborn, J. (1997). Constructivist metaphors of learning science. *Science & Education*, Special double issue on philosophy and constructivism, *6*(1–2), 121–133. https://doi.org/10.1023/A:1008642412858.
Ogborn, J. (1999). Unpublished MS privately communicated, based on Ogborn, J. (1994). Theoretical and empirical investigations of the nature of scientific and common sense knowledge. PhD Thesis of the University of London, May 1994.
Papadouris, N., Vokos, S., & Constantinou, C. P. (2018). The pursuit of a "better" explanation as an organizing framework for science teaching and learning. *Science Education, 102*(2), 219–237.
Posner, G. J., Strike, K. A., Hewson, P. W., & Gerzog, W. A. (1982). Accommodation of a scientific conception: Toward a theory of conceptual change. *Science Education, 66*, 211–227.
Thagard, P. (2008). Explanatory coherence. In J. E. Adler & L. J. Rips (Eds.), *Reasoning* (pp. 471–513). Cambridge: Cambridge University Press.

Appendix B. A Class Session on Thermal Insulation

This Appendix details a case (previously analysed in Viennot 2013) where a document intended for teachers in lower secondary education shows an internal contradiction in physics. It seems that this has not been detected by the authors of the document, since it is neither recorded nor commented on. This example is provided to support an existence theorem: it is not totally implausible that one encounters, especially during teaching practice, an internal contradiction in physics that is not identified as such.

This class session is proposed in a DVD to promote a type of education that emphasizes inquiry. The presentation is made through a series of episodes, in which various groups are observed, supervised by three teachers (Life and Earth sciences, Physics and Chemistry, and Technology) working in turn. In this composite sequence, each episode illustrates a phase of the process (written in italics on the DVD, as below). After the phase of *Identification of the problem giving rise to the research* (we are on Mars and it is cold), *students suggest strategies to deal with the problem.* On the board, a question is asked*: How to protect buildings and clothing from heat and cold?*

T (teacher) – So, to answer this question we will try to make a list of the materials we could use. So it's up to you to make proposals now.

After this encouragement of the teacher comes a phase called *Collecting ideas and experience protocols.*

S (*a student*) -Survival blankets
T- Survival blankets. So, do you know what the material is? What's on a survival blanket that allows you to survive?
S- Aluminium
T- Yes? For some, yes. So now we're going to list the materials. I'm going to write on the board and then we'll just take some of those materials because we don't have....

The core of the sequence then consists of material proposals by the students, *hypothesis* making, setting up of experiments, all with very active students involved in the discussion. For example, containers the size of a cup are filled with hot water and covered with lids made of various materials. An ice cube is placed on the lid of each container and the time it takes for the ice cubes to melt is compared. In another episode, the experiment is done by placing the sheet of material used for the lids directly on the work table: in both cases, it is with aluminium that the ice cube melts the fastest. This is a crucial step in a teacher-student exchange:

T- So, here it melts faster, it means that aluminium (back to the original hypothesis). Is it a good insulator?
S- No.
T- No. Can it protect us from the cold?
(Chorus students)- No

The analysis presented here focuses on two observations: i) in none of the experiments presented in the DVD there is mention of the geometric dimensions of the material tested as factors to be taken into account; ii) the last episode quoted here presents two conclusions which seem accepted by all and can be formulated as follows: aluminium is not a good insulator, and it cannot protect us against the cold. The first statement is confirmed later, the second is nowhere in the DVD related to the existence of survival blankets containing aluminium.

This conclusion – one cannot protect oneself from the cold with aluminium – is problematic because it is in flagrant contradiction with what was accepted at the beginning of the sequence: the beneficial use of survival blankets. In fact, a physical analysis of the functioning of a survival blanket should consider several energy transfer processes: conductive, convective, radiative. Instead, the conclusion expressed after the experimentation took only conduction into account, whereas the benefits of survival blankets are due to their radiative properties.

Because of the structure of the DVD, it is not known whether the same students confronted these two contradictory ideas – *survival blankets allow you to survive* and *aluminium cannot protect us against the cold*. But the teacher who watches this DVD as part of her/his training, will not know if such a discussion needs to take place or how to conduct it. The DVD implies that there is no problem.

The curious reader will find in Appendix H a physical analysis of the functioning of a survival blanket (see also Viennot and Décamp 2016a). This analysis takes into account several energy transfer processes (conductive, convective, radiative) whereas the conclusion of the discussion reported in this DVD only takes conduction into account.

Reference of Appendix B

Viennot, L. (2013). Les promesses de l'Enseignement Intégré de Science et Technologie (EIST) de la fausse monnaie? *Spirale* n° 52, 51–68. (The Promises of Integrated Science and Technology Education (ISTE): Fake Currency?).

Appendix C. Atmospheric Composition and Radiocarbon Dating

The subject of radiocarbon dating seems to be well known to future physics teachers. It is based on the radioactive decay of atoms of one isotope of carbon ($^{14}C \rightarrow {}^{14}N$ + electron + antineutrino). The law of radioactive decay gives the number N(t) of ^{14}C atoms of an initial population N_0 remaining after a duration t: $N(t) = N_0 \exp(-\lambda t)$. The value of the constant λ is linked to time τ (the "half-life", here 5730 years) at the end of which only half of the initial atoms remain: $\lambda = \ln 2/\tau$. For a given initial population N_0, therefore, there are fewer and fewer radioactive atoms: a measurement of the [$^{14}C/^{12}C$] ratio in a dead organism will make it possible to evaluate the time over which the atoms of the initial population have decayed (^{12}C atoms are stable). In any case, in order to know the value of the exponential and thus the time elapsed since the death of the organism, it is necessary to know the composition [$^{14}C/^{12}C$] of the carbon skeleton at that date. However, it is admitted that as long as it is alive, the [$^{14}C/^{12}C$] proportion of a living organism's skeleton is the same as that of the atmosphere because of the exchanges linked to its metabolism. On the other hand, once death has occurred, these exchanges cease and therefore nothing comes to replenish the ^{14}C population in the skeleton. The question then is how to know the [$^{14}C/^{12}C$] composition of the atmosphere at the time of the organism's death. In response to this question, it is accepted that the composition of the atmosphere is the same as at present. Could it be that in the atmosphere the radiocarbon does not disintegrate? This is absurd! There is only one possibility: a mechanism adds radiocarbon atoms to the atmosphere at the same rate as they disappear through disintegration. This mechanism is linked to the action on nuclei of nitrogen (^{14}N) of neutrons called "cosmic" (actually resulting from the action of cosmic particles on nuclei of atmospheric atoms, e.g. oxygen): this action generates radiocarbon atoms (neutron $+^{14}N \rightarrow {}^{14}C$ + proton). Very well, but how is it that the production thus achieved balances exactly radioactive decay? And how was a given value of the $^{14}C/^{12}C$ composition generated and stabilized?

To facilitate the understanding of these points, we have proposed the analogy below (Décamp and Viennot 2015).

Analogy of Removals

Imagine a country with a stable total population, N_{Tot}. The population is divided into two categories: rural (N_R, e.g. ^{14}C) and urban (N_U, e. g. ^{14}N) populations. Suppose that each year 10% of urban dwellers (from N_U number) move to the countryside and 40% of rural dwellers (from N_R number) move to the city.

Can the population of each category be stable? Yes, if they have values such that the exchanges are balanced, i. e.:

$$0{,}01 N_U = 0{,}04 N_R$$

or

$$N_U = 4/5 \mathrm{N}_{\mathrm{Tot}} \text{ and } N_R = 1/5 \mathrm{N}_{\mathrm{Tot}}$$

If the urban population is lower than this equilibrium value, there will be more moving from the countryside to the city than in the opposite direction, and therefore the urban population will grow to its equilibrium value. If the urban population is higher than this equilibrium value, there will be more moves from the city to the countryside than the other way around, and therefore the urban population will decrease to its equilibrium value.

Thus, whatever the initial distribution of the population between urban and rural areas, it is the same proportion of inhabitants who will, in the end, settle between urban and rural housing.

This intriguing result is, in fact, due to the multiplicative structure of the temporal rates of change of residence: these are proportional to the existing populations.

This analogy lends itself well to the understanding of radiocarbon dating, as follows:

Table C.1 Elements of questioning and knowledge about carbon-14 dating and the academic level necessary a priori to express them, and possibly answer them

Question to ask ourselves	Knowledge element required[a] and accessible at the end of secondary school	(Additional) knowledge element required[a] and accessible at university level before graduation
How is the relative concentration [$^{14}C/^{12}C$] of the studied organism at the time of death found?	Carbon isotopes ^{12}C, ^{14}C	
	Food chains of living organisms, exchanges between plants and the atmosphere.	
What was the composition of the atmosphere [$^{14}C/^{12}C$] at the time of the organism's death?	*At the time of his death, an organism has the same ratio [$^{14}C/^{12}C$] as that found in the atmosphere at that time.*	
	The composition of the atmosphere at the time of the organism's death was the same as it is today.	
If we accept the stability over time of the relative composition [$^{14}C/^{12}C$] in the atmosphere, how can we justify this assertion?	Radioactive decay: The number N(t) of ^{14}C atoms in an initial population N_0 remaining after a time t is $N(t) = N_0 \exp.(-\lambda t)$	
	Nitrogen atom, ^{14}N: Composition of the nucleus. The disintegration of a ^{14}C atom produces a ^{14}N atom and the nuclear reaction of a ^{14}N atom with a neutron produces a ^{14}C atom.	
	The time rate of each of these transformations is proportional to the existing population concerned.	Not strictly necessary:
	The total number of ^{14}N + ^{14}C atoms remains constant over time.	neutron + $^{14}N \rightarrow$ ^{14}C + proton
	See the text for the analogy of removals.	$^{14}C \rightarrow$ ^{14}N + électron + anti neutrino

[a]Required: at least to ask the question, and possibly answer it

When a radiocarbon atom (^{14}C) disintegrates, it produces (among other things) a nitrogen atom (^{14}N). When a nitrogen atom (^{14}N) reacts with a cosmic neutron, it produces (among other things) a radiocarbon atom (^{14}C). Everything happens as when two populations exchange individuals (who "move"). In addition, the proportionality of exchange rates to existing populations ensures, as in the case of rural and urban populations, the stabilisation of their numbers after a transitional phase, regardless of the initial situation or after accidental situations (for example, for radiocarbon, a volcanic eruption).

Table C.1 summarizes the elements of questioning and knowledge about carbon-14 dating as well as the level necessary a priori to express them, and possibly answer them.

References of Appendix C

Décamp, N., & Viennot, L. (2015). Co-development of conceptual understanding and critical attitude: Analysing texts on radiocarbon dating. *International Journal of Science Education, 37*, 2038–2063.

Appendix D. Magus Effect for Non-specialists

This example is still a case in which, without necessarily understanding everything about the situation mentioned, we can identify contestable aspects of the proposed reasoning, here a strong incompleteness of the demonstration and an indirect denial of a law. The topic is the non-parabolic trajectory of a ball spinning on itself (Magnus effect, Fig. D1). The author of a popularization document for college teachers explains:

> … we know that pressure and speed are linked. Imagine that there is a pressure difference between two adjacent points in the air. Because of this pressure difference, the air is undergoing a pressure force that tends to accelerate it to the point where the pressure is lower; in a way, the air is pushed by the high pressure. In other words, the speed increases in the lower pressure region. We can therefore sum up by saying that the speed is higher where the pressure is lower and, in an equivalent way, that the speed is lower where the pressure is higher.

This text is supported by Fig. D2. This argument seems to refer implicitly to Bernoulli's theorem, i.e.: When the flow of an inviscid and incompressible fluid (of density ρ) is steady, and g being the gravitational body force per unit mass, the quantity $v^2/2 + p/\rho + g\,z$ (density: ρ ; pressure: p; flow speed v; altitude: z) is constant along a stream line (if $curl\,\vec{v} = \vec{0}$, this quantity is the same throughout the fluid). Note that here no application conditions are mentioned. How do we know now where the speed is *lower*? Lower than what, by the way? The text above suggests a comparison between speeds of the same air mass at different times: "the air is pushed (…) the speed increases". In support of this discussion, a drawing shows two adjacent stream tubes with parallel streamlines as well as two sections with different fluid velocities in one of the tubes. With parallel streamlines, the speeds in question cannot be different for an incompressible fluid. Secondly, the speeds and thus the pressures that we compare are those of two points (above and below the ball) that are not on the trajectory of the same air mass: how could we understand?

Table D.1 summarizes the elements of questioning and knowledge about the text analysed here as well as the academic level which is necessary a priori to express them and possibly answer them.

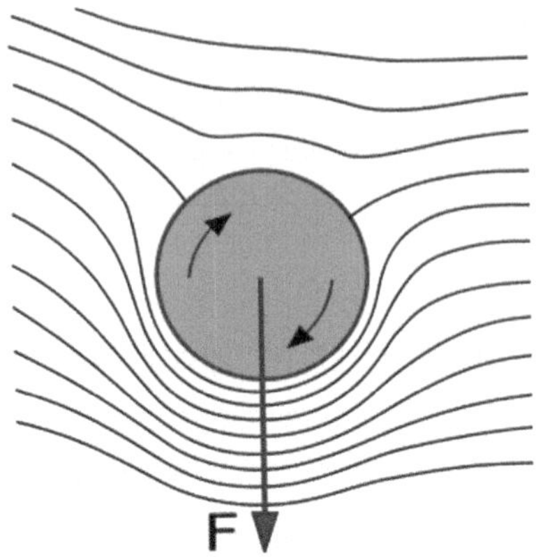

Fig. D1 Magnus effect, for a fluid moving (from right to left) relatively to a rotating ball: the part of the ball in contact with narrower stream tubes is subjected to lower pressure forces

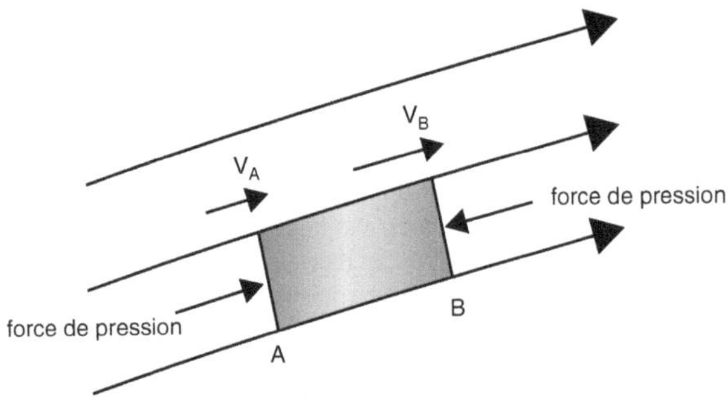

Fig. D2 A figure, in a popularization text, to explain the mutual dependence between fluid speed and pressure

Table D.1 Elements of questioning and knowledge about the text analysed in this Appendix and the level necessary a priori to express them and possibly answer them

Question to ask ourselves	Knowledge element required[a] and accessible at the end of secondary school	Additional knowledge element required[a] and accessible at university level before graduation
How can an incompressible fluid move at different speeds (V_A, V_B) across two cross sections (A, B) in a tube of constant cross-section?	Cylindrical stream tube: cross-sections of same area.	Incompressible fluid Flow rate Steady flow *Here, the tube is of constant cross-section:* *If the fluid is incompressible, the velocity of the fluid remains constant along the tube* $\vec{\nabla}\cdot\vec{v} = 0$ *So the explanation proposed here is not valid.*
Which object is concerned by the Newtonian balance implicitly used?	Newton's second law *Here, the object considered is the fluid located in a stream tube between two cross-sections of this tube.*	

(continued)

Table D.1 (continued)

Question to ask ourselves	Knowledge element required[a] and accessible at the end of secondary school	Additional knowledge element required[a] and accessible at university level before graduation
Have all the forces involved been taken into account?		Viscosity, Zero viscosity liquid (inviscid fluid) *Here, it is not mentioned that the proposed reasoning is not valid in case of non-zero viscosity.*
How does the comparison of speeds along a given stream tube explain a pressure difference between different points on the ball surface?	*If the explanation is accompanied by a drawing showing the air stream lines around the ball (as in Figure 30), it becomes clear that not all the different points of the ball are affected by the same stream tube. This clarification is not given here.*	As for the relationship between pressure and velocity, Bernoulli's theorem (see text) solves the question for an incompressible and inviscid fluid in a stationary regime. *To achieve a consistent explanation, it must be taken into account that, in all stream tubes, the pressure is assumed to be the same far from the obstacle and that it is more or less decreased in contact with the ball depending on the speed of the air relative to the ball – relative speed which depends on its rotation.*

[a]Required: at least to ask the question, and possibly answer it

Appendix E. The Pierced Bottle and the Range of the Jets

This Appendix draws widely on Viennot 2014, 135–139.

It is common to state that the range of water jets emerging from a pierced bottle and impacting a support at the bottom of the bottle increases with the distance between the hole and the free surface. That suggests (some would say, demonstrates) that at each hole the pressure increases with depth. It is clear that such a conclusion would show a lack of vigilance, coupled with confusion between the accuracy of a conclusion (pressure increases...) and the value of a demonstration. The very principle of this demonstration is contentious (see the calculations below): we want to show the increase in what is often called "hydrostatic" pressure even though we are dealing with a dynamic situation. The standard formal treatment of the situation, detailed below, is to calculate the exit velocity as if the water had fallen from the surface in free fall, at constant pressure – a supreme irony. It is therefore difficult to link the jet range to an outlet pressure, so another variable is chosen: the water exit velocity, which is horizontal. This depends on (the square of) the depth of the hole. Except that, if we actually perform the experiment (it is preferable to use an overflow system to maintain a steady total water height), we observe a maximum range for the mid-height hole (Chap. 2, Fig. 2.8), and equal ranges for holes symmetric about the mid-hole. This is due to the fact that the exit speed of the water is not the only relevant variable. The duration of free fall out of the bottle also contributes to the horizontal distance moved by the water before impact. It is finally the product of these two factors which accounts for the observed impacts. Exit velocity and time of free fall are respectively linked (via their square) to the distance from a hole to free surface and to its distance to the support. The product of these quantities accounts for the position of the impacts.

Standard Calculation

The classic approach is to use Bernoulli's theorem for this situation (Acheson 1990, 10). This presupposes that the flow regime is steady and that the liquid is incompressible and inviscid. The application of this theorem to two points on a streamline (Fig. E.1) located respectively on the free surface (A) and at the water outlet (B, at height h) leads to:

$$v_A^2/2 + p_A/\rho_{water} + g\,z_A = v_h^2/2 + p_B/\rho_{water} + g\,z_B \tag{1}$$

We then assume that the flow speed v_A at the free surface (whose area is much greater than that of the hole) is virtually zero compared with the exit flow speed v_h from the hole being considered. The water pressure is the same at the free surface as at the exit hole, i.e. atmospheric. Eq. (1) is then written.

$$p_{atm}/\rho_{water} + g\,z_A = v_h^2/2 + p_{atm}/\rho_{water} + g\,z_B$$

Therefore, the square of the speed, v_h^2, is proportional to the difference z_A- z_B, i.e. the difference between the height of the water surface, H and that of the hole, h:

$$v_h^2 = 2g(H-h) \tag{2}$$

It is interesting to note that this formula is the same as that obtained with the free fall model. The relationship p_h-p_{atm} = $\rho g(H\text{-}h)$, where p_h-p_{atm} is the difference in pressure between the altitudes h and H, cannot be used because it assumes

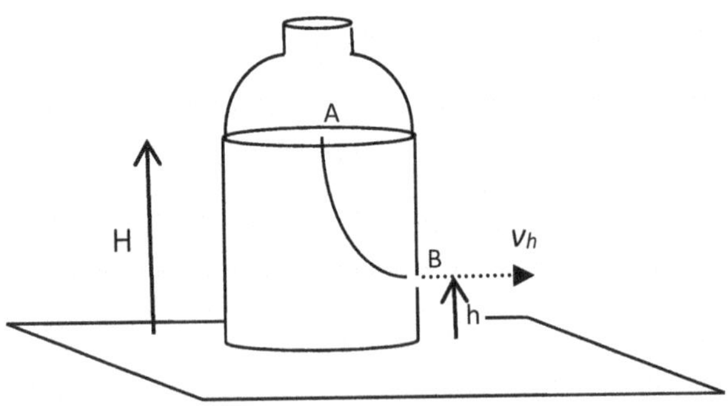

Fig. E.1 Application of Bernoulli's theorem at two points on a streamline for the flow of water within the bottle

hydrostatic conditions which clearly do not hold in the case of an accelerating fluid stream.

Another quantity affects the range of the water-jet: the time t_{ff} of the water's free fall outside the bottle, such that $h = (1/2)g\, t_{ff}^2$. The square of this time, t_{ff}^2, is proportional to the altitude h of the hole as measured from the table onto which the jet falls:

$$t_{ff}^2 = 2h / g \tag{3}$$

Assuming the water jet is horizontal at the exit hole, its range is obtained by multiplying the velocity v_h at the exit hole by the time of free fall, t_{ff}.

$$d = v_h\, t_{ff} = 2\left[h(H-h)\right]^{1/2} \tag{4}$$

The range of the jet therefore depends on the product of the two distances whose sum is H. With a stable level of water in the bottle, a steady flow regime is observed where the maximum range is obtained for the mid-height hole:

$$h_m = H/2.$$

In this model, two holes symmetric about h_m will give rise to jets of the same range.

It should be made clear that practical realisation of this experiment confirms the predicted sequence in the observed water jet ranges, but not their numerical values. Questions remain about the holes as well as about the viscosity of the water. For further details, see (Planinšič et al. 2011).

Table D.2 summarizes useful questions and knowledge about the documents on the pierced bottle discussed in this Appendix, as well as the academic level needed a priori to express them, and possibly answer them.

Table D.2 Elements of questioning and knowledge about common documents on the pierced bottle

Question to ask ourselves	Knowledge element required[a] and accessible at the end of secondary school	Additional knowledge element required[a] and accessible at university level before graduation
According to the documents quoted in Fig. 2.8, a hole at the level of the support would result in a greater range, but the water, accelerated by its weight in the vertical direction, would immediately hit the support.	Free fall, time equations in horizontal and vertical directions.	
	It is necessary to take into account the time of free fall outside the bottle, not only the exit speed.	Condition of independence of horizontal and vertical movements: absence of viscosity and friction of water on air

(continued)

Table D.2 (continued)

Question to ask ourselves	Knowledge element required[a] and accessible at the end of secondary school	Additional knowledge element required[a] and accessible at university level before graduation
	Energy conservation (no dissipation) for a drop that passes through an incompressible and inviscid fluid *(here under the effect of gravity).*	Viscosity, Inviscid fluid
		The energy balance can be presented as an application of Bernoulli's theorem.
	Extremum of the function $d(h) = 2[h(H\text{-}h)]^{1/2}$	
	The derivative of this function is zero for h = H/2	*Here, it is not mentioned that the proposed reasoning is not valid in the case of non-zero viscosity (this is all the more the case when the water is cold), or of appreciable friction on the air (this is all the more the case when the path of water in the air is long).*
What is the relationship between this analysis and the increase in hydrostatic pressure with depth?	Hydrostatic pressure	
	The pressure in the water at equilibrium is not in question here. We are only talking about water in accelerated fall.	
It is useful (and unusual) to point out that the treatment of this problem via Bernoulli's theorem is nothing but the expression of an absence of dissipation in water. Not mentioning the conditions of application of Bernoulli's theorem further blurs this idea. It is common to forget the role of the time of free fall and to consider only the exit speed of the drop.		

[a]Required: at least to ask the question, and possibly answer it

References of Appendix E

Acheson, D. J. (1990). *Elementary fluid dynamics*. Oxford: Clarendon Press.

Planinšič, G., Ucke, C., & Viennot, L. (2011). *Holes in a bottle filled with water: Which water jet has the largest range?* Muse project of the EPS-PED: http://education.epsdivisions.org/muse

Viennot, L. (2014). *Thinking in physics the pleasure of reasoning and understanding*. Dordrecht: Springer.

Appendix F. Batteries, Electrolytic Cells and Direction of Current

In the first years at university, electric circuits is a topic sometimes developed after electrostatics and magnetostatics and treated as a special case of electromagnetism. While some presentations of this kind are well conducted and consistent, a "simplified" version of this approach in high school can pose problems of coherence and completeness. Thus, we often find the following explanation, based on electrostatics, to explain the direction of the current (Academy of Bordeaux, n.d.):

> Between the two terminals of a battery there is a permanent difference in the density of the free electrons: the negative terminal has a higher than normal electron concentration while the positive terminal is deficient in electrons. If an electrical circuit is connected to the battery, the free electrons of the circuit are attracted by the positive terminal, repelled by the negative terminal of the battery. They circulate from the negative terminal to the positive terminal outside the generator.

None of the sentences in this text violates the laws of electrostatics, but the last sentence (*outside the generator*) invites us to wonder what is going on inside the battery and to question this simplified explanation. Indeed, if a higher concentration of electrons creates a charge singularity, shouldn't it attract isotropically the positive charges? And if so, should we understand that the direction of current inside the battery is contrary to the direction of current outside the battery? A well-informed physicist knows that this is not the case, but at the same time, they will look with caution at this type of explanation involving only two specific charge singularities.

However, this type of explanation can be found again in the study of electrolytic cells (Academy of Bordeaux, n.d.): "As an ionic solution is liquid, all the particles that make it up move freely (property of the liquid state) and randomly. When this solution completes an electrical circuit, all the particles carrying electric charges (the ions) no longer move freely, but will be attracted by the generator terminals:

> The anions (bearing negative charges like electrons) are attracted by the + terminal of the generator[...].
>
> The cations (bearing positive charges) are attracted by the – terminal of the generator. [...].

Criticism is made easier here by the fact that ions are explicitly mentioned, ions which are known to be involved in conduction inside the battery: It is intriguing that they are attracted by a terminal of the generator when they are in the electrolytic cell but repelled by the same terminal when they are in the battery (Fig. F.1).

Other formulations, in terms of surface charge distribution, make it possible to improve these explanations (Fig. F.2). This is the case, for example, of the presentation adopted by Chabay and Sherwood (2006). Here, this distribution explains the field inside a conductor and therefore the direction of the current.

Unfortunately, as the electric field circulation along a closed circuit is zero ($\vec{\nabla} \wedge \vec{E} = \vec{0}$ in direct constant current), the electrical potential along an electric circuit necessarily has a maximum and a minimum (unless it is uniform). Our initial

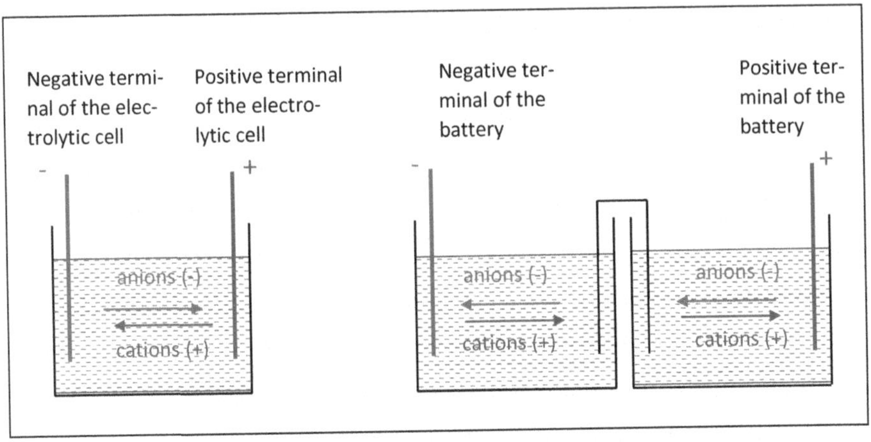

Fig. F.1 The migration of ions in an electrolytic cell is often explained by the fact that the (positively charged) cations are attracted by the negative electrode of the electrolytic cell and that the (negatively charged) anions are attracted by the positive electrode of the electrolytic cell. It is intriguing to see that the migration of ions in a battery is in the opposite direction

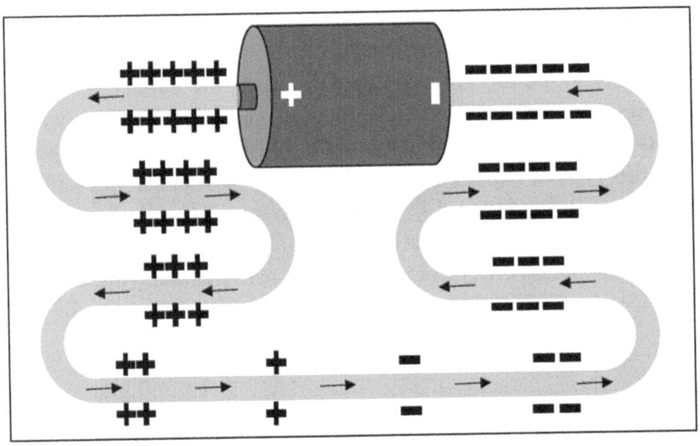

Fig. F.2 (repeat of Fig. 2.10): Charge distribution in a simple electric circuit in a quasi-stationary regime. Surface charges are represented at different points of the circuit so that the charge gradient and the resulting field are explicit. See in particular Chabay and Sherwood (2006). Not shown here, the charges located on the curved parts of the circuit are essential to explain that electrons follow (on average) the wires

question can then be expressed in terms of potentials: if the electrons go in the direction of the potential gradient outside the battery (and possible positive charges go in the opposite direction), can the anions go in the opposite direction to the potential gradient inside the battery (unlike cations) (Fig. F.1)? Overall and seen from the outside, it seems that “in a generator,[...] the current[...] circulates in the direction of increasing potential” (Université de Compiègne n.d.).

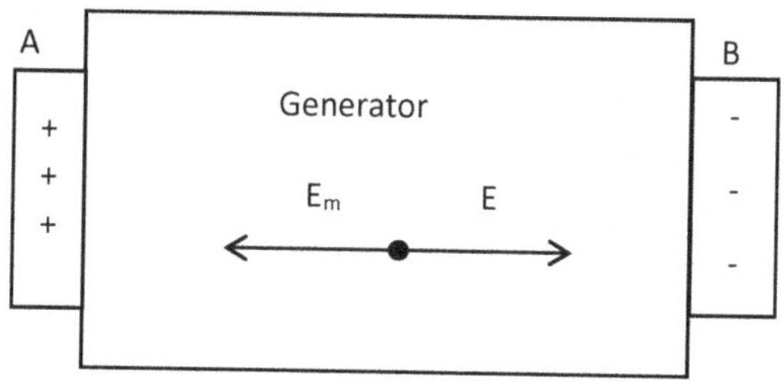

Fig. F.3 Figure illustrating the electric field and the electromotive field within a generator (Bruhat 1963)

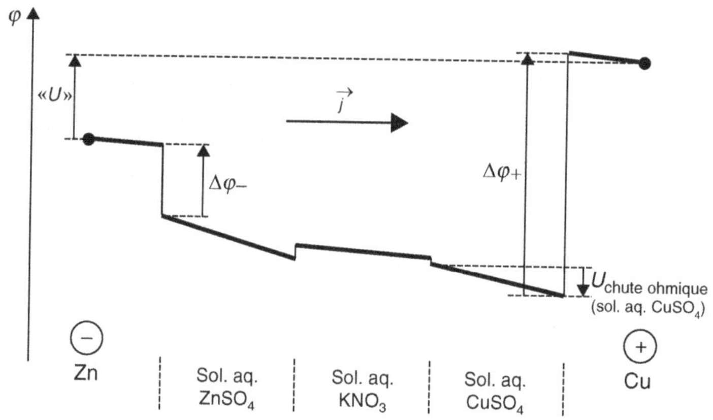

Fig. F.4 Potential profile in a Daniell electrochemical cell (Lefrou et al. 2012, p. 73). U is the potential difference between the terminals of the cell, the terms $\Delta\varphi^-$ and $\Delta\varphi^+$ represent the potential drops between each electrode and the liquid. $\vec{j}$ is the current density

To avoid this paradox, different proposals are available. The most traditional is to postulate the existence of a second field called "electromotive". We find trace of it in this extract from Bruhat (1963): "The essential characteristic of a generator is to create, by various mechanisms, an electromotive field" (p. 251). The figure accompanying this excerpt is explicit (Fig. F.3).

The raison d'être and the interaction associated with this electromotive field are not mentioned.

Other authors, cautiously, move the problem away from the domain of dynamics (or even physics) and simply talk in terms of energy: on this occasion, it is often mentioned that chemical energy is converted into electrical energy without any clear definition of what these terms refer to. Another approach is to use a hydraulic analogy. Its proponents, as well as being well aware of the problem, allude to a "pump" which allows the water in the circuit to "rise" without, again, the physical mechanism of this "pump" being precisely identified. However, a more detailed examination of the situation makes it possible to remove ambiguities and convince oneself that the ions circulating inside the generator are moving in accordance with the laws of electrostatics, as illustrated in Fig. F.4.

Table F.1 Elements of questioning and knowledge useful to discuss common documents on the circulation of charges within a battery

Question to ask ourselves	Knowledge element required[a] and accessible at the end of secondary school	Additional knowledge element required[a] and accessible at university level before graduation
In which direction and why do charges circulate inside a battery?	*Positive charges are attracted by negative charges and reciprocally.*	$\vec{E}$ electric field
		V electric potential
		At a point where there is an electric field $\vec{E}$, the positive (or negative) electric charges are subjected (among other things) to a force which is of same direction as (or opposite to) this field. In a medium of conductivity the density of current at this point is: $\vec{j} = \sigma \vec{E}$
		Along a field line, the electric field is directed towards the decreasing potentials.
		Oxidation potential
		$\vec{E} = -\overrightarrow{grad}\,V$
Teaching rituals: In a closed circuit, a reasoning based on the attraction/repulsion by the charges on the ends is used to account for what happens outside the battery, a reasoning that is abandoned inside the battery: In this case, the notions that dominate or have dominated the explanation of what happens inside the battery are: - an *electromotive* field - the values of the oxido-reduction couples in play.		

[a]Required: at least to ask the question, and possibly answer it

Figure F.4 shows that there are two important potential jumps. The fact that these two jumps are not of the same size explains the potential difference between the terminals of the cell on open circuit ($E = \Delta\varphi^+ - \Delta\varphi^-$). In closed circuit, the potential difference U between the terminals is slightly smaller, given the potential drop inside the cell due to the current I (which is the same as outside the cell) through the internal resistance r: $U = E\text{-}rI$ (where $E = \Delta\varphi^+ - \Delta\varphi^-$).

Table F.1 summarizes elements of questioning and knowledge useful to discuss common documents on the circulation of charges within a battery, as well as the level needed a priori to express them, and possibly answer them.

References of Appendix F

Academy of Bordeaux. http://webetab.ac-bordeaux.fr/Pedagogie/Physique/Physico/Electro/e02coura.htm

Academy of Nancy. http://www4.ac-nancy-metz.fr/clg-cassin-eloyes/Disciplines/Site_physique_chimie/Troisieme/Chapitre_C3/courant_ion.html

Bruhat, G. (1963). Electricité, p. 251 and fig. 17, Paris: Masson & Cie.

Chabay, R. W., & Sherwood, B. A. (2006). Restructuring the introductory electricity and magnetism course. *American Journal of Physics*, *74*, 329–336.

Lefrou, C., Fabry, P., & Poignet, J.-C. (2012). *Electrochemistry the basics,* with examples. Berlin/Heidelberg: Springer-Verlag (fig. 2.11, p. 73).

Appendix G. Capillary Rise and the Force That 'Lifts' the Liquid

This appendix details the problems posed by a typical analysis of the equilibrium involved in capillary phenomena and provides a more satisfactory explanation.

Figure G.1 shows a common pattern for the analysis of capillary rise (often water in a glass tube) and for the calculation of the contact angle θ between liquid (L) and solid (S) in the presence of gas (G).

The coefficients (γ_{LG}, γ_{SL} and γ_{SG}) refer to the forces per unit length drawn at the interfaces concerned, respectively. Thomas Young establishes in 1805 the formula linking these coefficients and the angle of contact θ:

$$\gamma_{LG}\cos\theta + \gamma_{SL} - \gamma_{SG} = 0 \tag{1}$$

A thermodynamic approach of this situation was proposed in 1878 by Gibbs. In this approach, the coefficients take on the meaning of free energy per unit of interface surface area at constant temperature. The equilibrium state corresponds to a minimum of this energy and therefore to a zero value of its variation with an infinitesimal displacement of the contact line. We can now interpret, for each interface, a term of the $\gamma\ dS$ type as the work done by a force $\gamma\overrightarrow{dl_1}$ moving through a displacement $\overrightarrow{dl_2}$ in its plane ($\overrightarrow{dS} = \overrightarrow{dl_1} \times \overrightarrow{dl_2}$). Then, to cancel the variation of free energy per unit surface area for a vertical displacement of an element of the contact line comes down to writing that the sum of the terms $-\gamma_{LG}\cos\theta\, dl_1$, $-\gamma_{SL}\, dl_1$, and $\gamma_{SG}\, dl_1$, is zero, as with relationship (1).

The resulting relationship resembles a force balance that would be zero in a vertical direction, as suggested by Fig. G.1, but the horizontal direction poses a problem from this point of view.

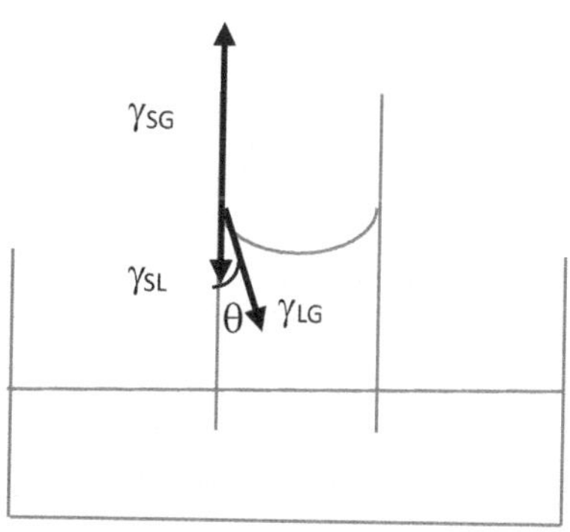

Fig. G.1 Capillary rise and contact angle between liquid (L) and solid (S) in the presence of gas (G); γ_{LG}, γ_{SG} are γ_{SL} are interfacial tensions (forces per unit length) for the liquid/gas, solid/gas and solid/liquid interfaces respectively, and θ is the contact angle

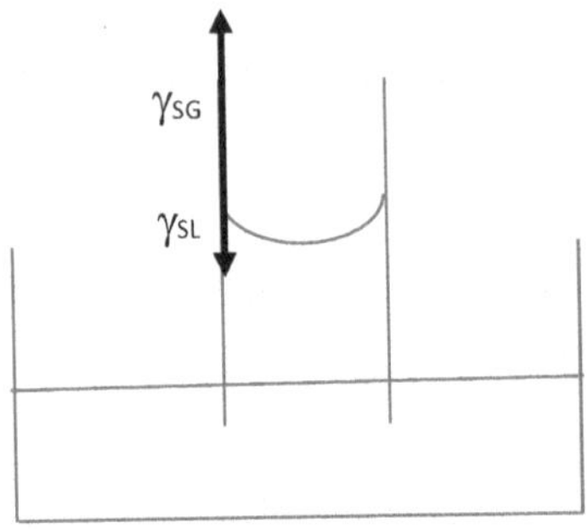

Fig. G.2 A common explanation of the capillary rise (notations as in the previous figure)

In addition, there is no way to connect these "forces" to the column height above the surface level in the tank. In fact, considering only two forces (Fig. G.2, see Viennot 2015) allows a calculation of the height of the liquid column, in line with this comment found in a popular book for physics teachers in lower secondary education: "The force that lifts the liquid is $\gamma_{SG} - \gamma_{SL}$ "(Quéré 2001, p. 158).

Finally, some authors (Berry 1971; Brown 1974; Das et al. 2011; Marchand et al. 2011) question the status of the forces – often qualified as "fictitious" – involved in these explanations, including the fact that they often seem to "pull" on an immaterial line. They analyse molecular interactions and their consequences in terms of localized forces around the interfaces, involving molecules on a typical thickness of two or three molecules. The domed shape of a water or mercury drop deposited on a horizontal support visually attests to the reality of these interactions and the associated surface tension. But from there to understanding the effects of these interactions tangentially on the surface, there is an important step forward.

To summarize the reasons for frustration with the documents described above (Figs. G.1 and G.2), there are at least four.

1. What looks like a balance of forces is not balanced in a horizontal direction.
2. These *"forces"* seem to act on lines (dimension: 1) and not on specified objects with a non-zero mass.
3. The text does not provide a means to account for the height of the water column above the surface in the container at equilibrium.
4. A smooth, vertical glass wall can only attract a molecule of water horizontally, for symmetry reasons (Das et al. 2011; Marchand et al. 2011). It is therefore impossible for a force involving such a glass wall to *lift* the liquid vertically.

One way to solve all these difficulties is to construct a dislocated diagram for the liquid corner that borders the free surface of the upper meniscus, the rest of the water in the column, the rest of the water in the container and the vertical glass wall (Fig. G.3). Taking into account the equilibrium of the liquid corner, the force modules per unit length of the air/water/glass contact line are respectively:

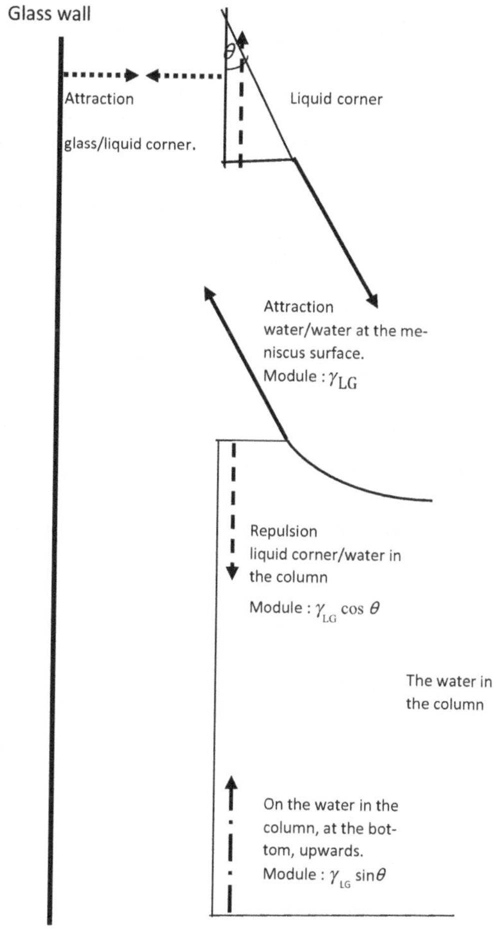

Fig. G.3 Capillary rise: 'dislocated' diagram concerning objects (liquid corner, glass and water in the column) defined by an element of the contact line between water, glass and air (perpendicular to the figure). The only complete free-body diagram (by unit length of the line of contact) is on the liquid corner. The weight of the liquid corner is neglected with respect to the other forces, the weight of the liquid column, *P*, is not represented (for a column of height *h*, $\boldsymbol{P} = \rho_{wat}\,\pi r^2_{tube} h\, g = \gamma_{LG} \cos\theta\, 2\pi r_{tube}$) and the horizontal and repulsive interaction between the water in the column and the glass wall is not shown. This diagram shows the zero force balance on the liquid corner and the overpressure in the water near the vertical wall. The role of atmospheric pressure is not represented: it exerts a normal force on each external surface (even immersed) of the various objects in play (liquid corner, liquid column, glass wall, …) whose total contribution to the Newtonian balances is zero. The only complete free-body diagram (by unit length of the line of contact) is on the liquid corner. Figure partly based on Das et al. 2011 and Marchand et al. 2011

Attractive interaction between glass and liquid corner, horizontal: $\gamma_{LG}\sin\theta$

Repulsive interaction liquid corner/the rest of the water in the column, vertical: γ_{SG},- $\gamma_{SL} = \gamma_{LG}\cos\theta$

Attractive interaction water/water, tangential to the free surface of water: γ_{LG} (vertical component: $\gamma_{LG}\cos\theta$, horizontal component: $\gamma_{LG}\sin\theta$).

The apparent contradictions pointed above disappear. The equilibrium of the water in a cylindrical column implies that the vertical force component $\gamma_{LG}\cos\theta$. $2\pi r$ is opposed to the weight of the water, $\rho_{wat}\,\pi r^2 h\, g$ (liquid cylinder radius: r, height: h, density: ρ_{wat}).

It follows that

$$h = 2\gamma_{LG}\cos\theta \,/\, \rho_{wat}\, g\, r$$

It can be observed that the height in question is inversely proportional to the radius of the tube, at given surface tension (Jurin's law).

It is particularly interesting to examine the paradox pointed out in Chap. 4. A smooth and vertical glass wall can only attract water molecules in a normal direction to the wall – here horizontal. How, then, can we speak of a "force that lifts the liquid"?

In fact, the difficulty stems from an overly local analysis of what causes capillary rise. In line with the habits of common reasoning (Chap. 3), attention is focused on what is visibly moving, i. e. the top of the column, and seeks a cause located in the same place. Hence this location at the very top of the liquid corner is commonly chosen for the forces that are supposed to explain the energy changes affecting the whole system. But nothing pulls the liquid up at this point. On the liquid corner itself, there is actually a thrust, which is provided by the liquid molecules compacted all along the wall due to the horizontal attraction they undergo. This analysis makes it possible to understand that at the other end of this compressed zone (here: below) a force is necessarily exerted (here: upward), which blocks the line of compressed molecules in one way or another. We can imagine a tube going down to the bottom of the vessel, which provides this blocking force. The discontinuity due to the lower orifice of the tube may also be responsible for an upward force, since the symmetry invoked above ceases to exist. From then on, the vertical equilibrium of the water column is conceived without particular difficulty, when all the vertical forces acting on it are taken into account, both at the bottom and at the top, including its weight.

The Centrifuge Analogy

An analogy can help us to understand these elements of analysis (Viennot 2015). It is a centrifuge (cylinder radius R) without a cover, partially filled with water and rotating (angular velocity ω). In the rotating frame of reference, a dynamic equilibrium implies a horizontal centrifugal force. This (inertial) force plays a role partially similar to that of the attraction of water molecules by glass, in that it *pushes* these

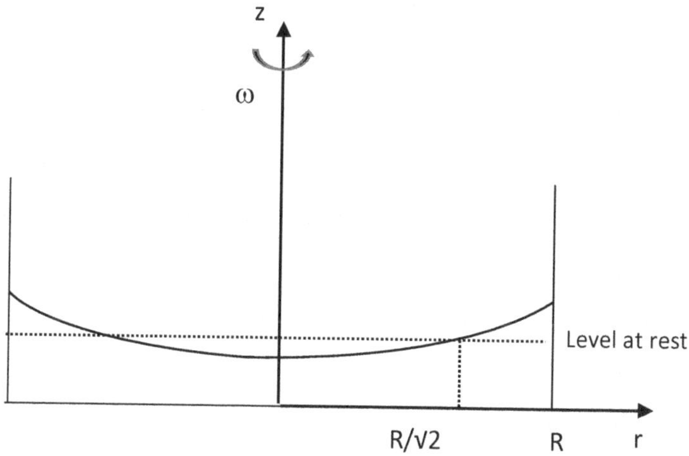

Fig. G.4 Dynamic equilibrium for a rotating fluid in a cylindrical centrifuge of radius R at constant rotation velocity ω

molecules horizontally towards the vertical wall against which they pile up. The free surface takes a paraboloid shape (Fig. G.4). Given the conservation of water volume, the centre of mass of the water is higher than at rest. A greater vertical force than the weight therefore acted on the water during the acceleration phase. Nothing drew the water from above, therefore it was pushed upward from below. It was the bottom of the vessel that exerted this additional repulsive force. Of course, the scale of the phenomenon is quite different from the case of a capillary rise: the compression of the rotating water concerns a large part of the vessel and not a zone with a few molecular diameters of thickness. But the main ingredients of the explanation are there: stacking of molecules piled up close to the wall because of a horizontal interaction with the glass and greater action of the support of the water on it, in a vertical direction, than at rest. In any case, we can clearly see from this analogy that it would be incongruous to analyze the situation of the centrifuge with a force that would lift the liquid by pulling it upwards.

With or without analogy, the analysis summarized in Fig. G.3 solves the problems raised in the four main critical arguments listed above. One might wish that these criticisms could be expressed by people who have not yet reached this clarification, but this is not the case in general, far from it (Viennot and Décamp 2018). Even after understanding the phenomenon and the explanatory value of the dislocated diagram, there is still, sometimes, a great resistance to questioning the diagram in Fig. G.1, as in this interview with a future teacher at the very end of her education:

Participant –[…] There (on the diagram in Fig. G.1), there are the three forces that start from the point of contact between the three, uh, so, uh, that's what's different, uh, from mechanics.

Interviewer – Do you mind if it's different?

- No, I'm not particularly shocked.
- You're not shocked, all right. It's not like mechanics, you tell me, but it doesn't matter?

- No, I don't mind.
- You would accept that there is a particular logic for capillarity which is not the logic of mechanics?
- Yeah, I wouldn't mind.
- Yeah, so there might be a type of balance just for capillarity?
- Maybe, maybe.

When the imperative of consistency is not perceived, the educator may find it difficult to encourage criticism. On the contrary, when this imperative is perceived and reasonably achieved, there are eloquent comments of satisfaction:

Interviewer – It is more complicated, you will confess (the diagram in Figure G.1 than the one in Fig. G.1)

Participant – Ah no!

- Ah no?
- Just because, there are more things doesn't mean that it's more complicated (...) No, no, it's not more complicated in the sense that we understand better.

Table G.1 summarizes elements of questioning and knowledge useful to discuss common documents on capillary rise, as well as the level needed a priori to express them, and possibly answer them.

Table G.1 Elements of questioning and knowledge useful to discuss common documents on capillary rise

Question to ask ourselves	Knowledge element required[a] and accessible at the end of secondary school	Additional knowledge element required[a] and accessible at university level before graduation
If the diagram in Figure 37 were to be interpreted as a free body diagram, why is this diagram not balanced in a horizontal direction?	Free body diagram	
	Second Newton's law	
	Here, it is not possible to interpret this diagram as a balanced free body diagram.	
If the diagrams in Figure 37 and Figure 38 were to be interpreted as a free body diagrams, on which system are the forces in question supposed to be exerted?	Free body diagram	
	Second Newton's law	
	Here, it is problematic to interpret these diagrams as concerning forces acting on a (mass) system. The term "interface", often used, suggests an immaterial line. It should be specified which molecules are concerned by these diagrams.	

(continued)

Table G.1 (continued)

Question to ask ourselves	Knowledge element required[a] and accessible at the end of secondary school	Additional knowledge element required[a] and accessible at university level before graduation
If the diagrams in Figure 37 and Figure 38 were to be interpreted as free body diagrams, how would they account for the existence of a water column in the tube?	Free body diagram	
	Second Newton's law	
	Here, it is problematic to interpret either scheme to justify the presence of the water column, since it is not specified on which system these forces are exerted.	A way to answer the questions asked is proposed in Figure 39. The forces per unit length of the contact line called *interfacial tensions* concern the molecules of each interacting body located near the separation surface, over a thickness of a few molecules.
	If these forces (by unit length of contact line) were supposed to be exerted on the water at its free surface, we observe that:	
How could the glass wall (homogeneous, smooth and vertical cylinder) exert a force other than horizontal on a water molecule?	*- the diagram in Figure 38 appears to be balanced in a vertical direction but does not mention the weight of the water concerned;* *- the diagram in Figure 38 has only two forces left (why?) and could be understood if something pulled the water upwards: it would be enough to add the weight of the column. However, a* glass wall (homogeneous, smooth and vertical cylinder) *can only exert a horizontal force on a water molecule.*	
How can the analogy of the open centrifuge (Figure 40) shed light on capillary ascension?	Free body diagram	
	Second Newton's law	
	The intuition of what happens in a centrifuge makes it possible to access the fact that the water rises on the edges without being drawn upwards at the level of its surface. So it is pushed from below.	
	Acceleration in a planar, horizontal, circular movement	
	The expression of the force exerted by the wall on a fluid element of mass dm ($d\vec{f} = -dm\,\omega^2\vec{r}$) and therefore of its reciprocal, is not even necessary. It is sufficient to understand that the acceleration of the fluid element is horizontal, and therefore also the wall-fluid interaction.	

[a]Required: at least to ask the question, and possibly answer it

References of Appendix G

Berry, M. V. (1971). The molecular mechanism of surface tension. *Physics Education*, *6*, 79–84.

Brown, R. C. (1974). The surface tension of liquids. *Contemporary Physics*, *15*(4), 301–327.

Das, S. Marchand, A., Andreotti, B., & Snoeijer, J. H. (2011). Elastic deformation due to tangential capillary forces. *Physics of Fluids*, *23*, 072006.

Marchand, A., Weijs, J. H., Snoeijer, J. H. & Andreotti, B. (2011). Why is surface tension parallel to the interface. *American Journal of Physics*, 999–1008.

Quéré, D. (2001). Bulles gouttes et perles liquides, In D. Jasmin, J. M. Bouchard, et P. Léna (Eds.), *Graines de science 3* (pp. 145–165). Paris: Le Pommier.

Viennot, L. (2015). Ascension capillaire Quand le verre semble "hisser" le liquide. *Bulletin de l'Union des Physiciens*, *977*, 1201–1212.

Viennot, L., & Décamp, N. (2018). The transition towards critique: Discussing capillary ascension with beginning teachers. *European Journal of Physics*, *39*, 045704. https://doi.org/10.1088/1361-6404/aab33f.

Young, T. (1805). An essay on the cohesion of fluids. *Philosophical Transactions of the Royal Society of London,* n° 95, 65–87.

Appendix H. Which Side to Put the Survival Blanket on?

Everyone knows that a survival blanket can help protect against the cold. Typically, it is an ultra-thin film (13 μm) of Mylar with one silver and the other gold. The common and recommended use of these products is to place the silver side towards you to protect yourself from the cold (towards the outside if it is necessary to protect yourself from heat). The reason given is that the silver side is more reflective and will therefore more effectively return the "heat" (as is often said) to the person at risk of hypothermia. In fact, a physical analysis of the situation in question (Viennot and Décamp 2016) does not make this choice obvious. For an opaque body, more reflectance means less absorbance and therefore less emissivity (see Table H.1). As can easily be seen from the experience shown in Fig. H.1, the emissivity of the silver side is lower than that of the gold side, which means that the reflectance of the silver face is higher than that of the gold face. One might wonder: for a blanket with only one silver side, wouldn't turning it inward lead to losing more energy by radiation towards the outside, because of the golden side? There is a dilemma.

A first step in understanding this physical situation is therefore to recognize that it escapes the obvious. Experimentation with teachers at the end of their studies (master's degree MEEF 2, Viennot and Décamp ibid.) shows how difficult it is for them to distance themselves from a common idea, reinforced by the "Red Cross" stamped instructions for use. As noted above, one in seven participants was able to

Table H.1 Relationships that model energy transfer by radiation (Griffiths 1999, see also Besson 2009)

For incoming radiation R, with a « spectral irradiance » $E_e(\nu)$ and « irradiance » (power per unit of area perpendicular to R) $E_e = \int_0^{\infty} E_e(\nu)\, d\nu$, the power *absorbed* per unit of area perpendicular to R is $\int a(\nu)\, E_e(\nu)\, d\nu = a_r\, E_e$

where a_r is the overall absortivity for all radiation R and is equal to the mean value, weighted by $E_e(\nu)$, of the absortivities $a(\nu)$ of all frequencies.

The total power *emitted* by a body at absolute temperature T per unit of area, or « radiant exitance», $M_e(T)$, is related to the radiant exitance of a black body at the same absolute temperature by the formula:

$$M_e(T) = \int_0^{\infty} e(\nu)\, M_e°(T, \nu)\, d\nu = e_t\, M_e°(T) = e_t\, \sigma T^4$$

σ is the Stefan constant. $M_e°(T, \nu)$ is the « spectral radiant exitance » of a black body at absolute temperature T. The overall emissivity, e_t, is defined as the ratio between the total power emitted per unit of area by a body at absolute temperature T and that of a black body at the same temperature. e_t is equal to the average value, weighted by $M_e°(T, \nu)$, of the emissivities of all frequencies.

$a(\nu) = e(\nu)$ for every ν; these coefficients can strongly vary with frequency.

Black body: $a(\nu) = e(\nu) = 1$ for every ν and $M_e°(T) = \sigma T^4$.

Fig. H.1 Samples of different materials glued to a metal kettle filled with boiling water: the temperatures displayed by a commercially available infrared radiometer are very different (survival blanket, silver side: 27 °C; semi-transparent adhesive tape: 95 °C; black adhesive tape: 96 °C; survival blanket, gold side: 38 °C). The radiometer displays the temperature that a black body emitting as much power as the target object per unit area would have. In practice, for an object of a given temperature, the temperature displayed by the radiometer is always lower than the temperature of the object, and this displayed temperature is all the greater as the emissivity of the object is higher"

use the experience in Fig. H.1 to challenge their initial idea without delay, even though they knew very little about the energy transfers involved (an "early critique" as defined in Chap. 5).

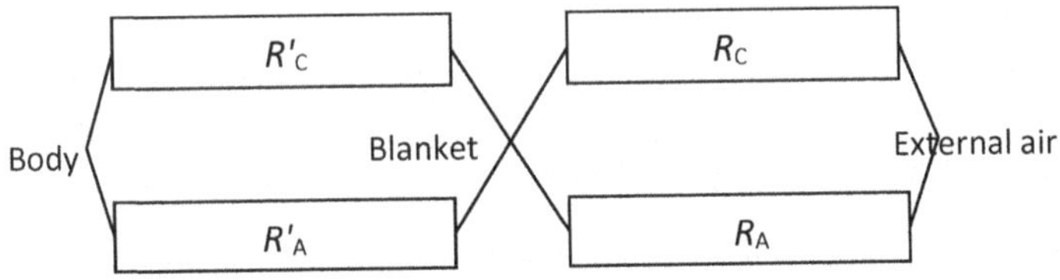

Fig. H.2 Electrical analogy for the transfer of energy between a person to be protected and the external environment

To go further, we can propose a simple model that distinguishes three objects – the person to be protected, the survival blanket and the outside air – each considered as having a single temperature (*Tp, Ts* and *Te* respectively). Between these objects, the net transfers envisaged are of two types: conductive-convective (coefficient *C*) and radiative (coefficient *A*); these occur in parallel. We consider the case where there is no direct solar flux on to the person to be protected. In linear approximation, the energy flow to the outside is then written

$$\Phi = (C' + A')(Tp - Ts) = (C + A)(Ts - Te)$$

where the coefficients C' and A' are relative to the zone between the person and the cover, while the coefficients C and A are relative to the zone (*boundary layer*) between the cover and the outside air. The radiative coefficients denoted A_a, A_o respectively concern the case where the silver or gold side is facing outwards.

All the transfer relationships (in this case energy) are then analogous to those that describe the electric charge flows (the currents) in a circuit schematized in Fig. H.2 according to potential differences (of which the equivalents here are temperature differences). It is then a question of comparing the equivalent coefficients of conduction (G: inverses of equivalent resistances) in the configuration "silver outside" and "gold outside".

One way to do this is to examine their difference:

$$G_a - G_o = \frac{(C' + A_o)(C + A_a)}{C' + A_o + C + A_a} - \frac{(C' + A_a)(C + A_o)}{C' + A_a + C + A_o}.$$

$$G_a - G_o = \frac{(C - C')(A_o - A_a)}{C' + A_o + C + A_a}$$

This is zero if $C = C'$, i. e. if the conduction coefficients for conductive-convective transfer are identical on each side of the blanket. Then the orientation of the blanket has no impact on the transfer. If $C \neq C'$, since $Ao > Aa$, the sign of $Ga\text{-}Go$ is the same as the sign of $C - C'$. In practice, in conditions favouring a high energy transfer towards the outside by conduction-convection ($C' < C$, wind and rain, $Go < Ga$), it is necessary to put the golden side outside. In other conditions ($C' > C$, dry and calm weather), the silver side must be outside (Fig. H.3).

Table H.2 summarizes elements of questioning and knowledge useful to discuss common documents on the survival blanket, as well as the level needed a priori to express them, and possibly answer them.

Fig. H.3 The best decision to protect ourselves against cold depends on the weather

Table H.2 Elements of questioning and knowledge useful to discuss common documents on the survival blanket

Question to ask ourselves	Knowledge element required[a] and accessible at the end of secondary school	Additional knowledge element required[a] and accessible at university level before graduation
Is it obvious that you have to turn the silver side of the blanket towards you to protect yourself from the cold?	For a first approach:	
	The silver side reflects more radiation (the 'heat') to the body (and emits less), but the gold side radiates more outwardly (reflects less).	
	Here, with only one silver side, it is necessary to find the best compromise between better reflecting thermal radiation towards oneself while wasting power towards the outside, or less reflecting thermal radiation towards oneself while reducing power dissipation towards the outside.	

(continued)

Table H.2 (continued)

Question to ask ourselves	Knowledge element required[a] and accessible at the end of secondary school	Additional knowledge element required[a] and accessible at university level before graduation
How to find the best compromise, when there is only one silver side?	Relationship between electrical potential (or here: The temperature difference), electrical (or here: Thermal) resistance and intensity (or here: Energy flow)	Action of a body on radiation. Reflection coefficient, *r*; transmission coefficient (zero for an opaque body), *t*; absorption coefficient, *a*.
	Equivalent resistance for dipoles in series or in parallel	Emissivity of a body.
		Box 9
		The model presented in the text and its electrical analogy solve the dilemma:
Common idea: To protect yourself from the cold, you have to turn the silver side of the blanket towards you, because that is how the "heat" is best reflected back to the body to be protected. This reasoning ignores the energy emitted to the outside by the golden side.		

[a]Required: to at least ask the question, and possibly answer it

References of Appendix H

Besson, U. (2009). Paradoxes of thermal radiation. *European Journal of Physics, 30*, 995–1007.

Griffiths, D. J. (1999). *Introduction to electrodynamics* (3e éd.). Upper Saddle River: Prentice-Hall.

Viennot, L., & Décamp, N. (2016). Co-development of conceptual understanding and critical attitude: Toward a systemic analysis of the survival blanket. *European Journal of Physics, 37*(1), 015702.

Appendix I. Water Pressure and Osmosis

The Basics About Osmosis

A prototypic situation for the osmosis phenomenon consists of two vessels separated by a semi-permeable membrane, i. e. one that can pass (in both directions) a solvent but not a solute dissolved in this solvent. Typically, osmosis occurs when solutions of different concentration are in contact with each side of a semi-permeable membrane as shown in Fig. I.1 (the membrane is located at the bottom of the U-tube).

Fig. I.1 An evocative representation of a situation of osmotic equilibrium: the small circles suggest the presence, in different concentrations on either side of the membrane, of solute in the solvent (which is shown only by its level in each branch)

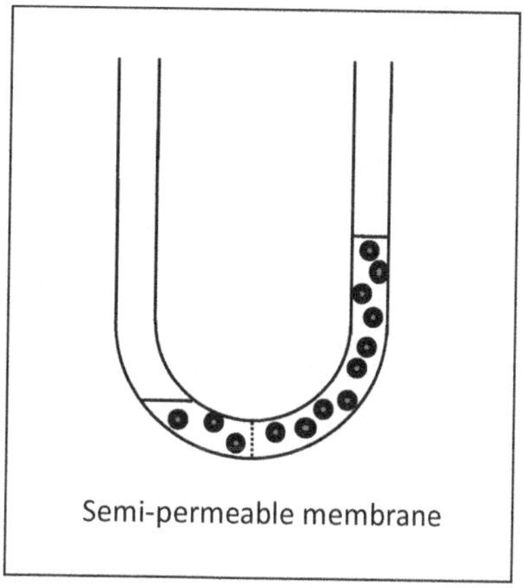

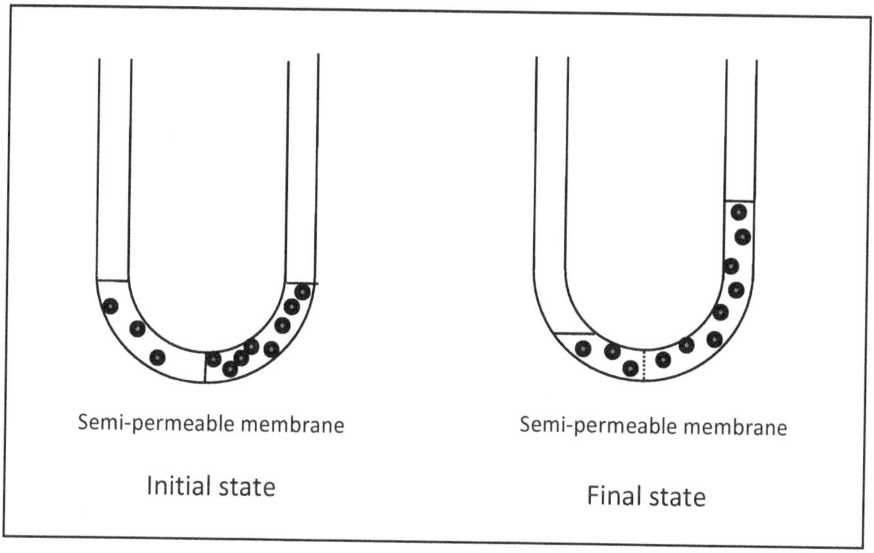

Fig. I.2 Repeat of Fig. 4.1, a type of drawing frequently observed (especially on Wikipedia) to introduce the phenomenon of osmosis, and which suggests that at equilibrium the concentration of solute is the same on both sides of the membrane

The equilibrium is achieved when a quantity relating to the solvent, in this case the "chemical potential of the solvent in the presence of solute", has the same value on both sides of the membrane. The chemical potential in question depends on the temperature, pressure and concentration (see Table 16 and Diu et al. 1989, 408–410). In the case of an (isothermal) equilibrium in the U-tube in Fig. I.2, there are therefore two possible cases: either the concentration of the solute and the liquid height in the branch are both identical on each side, or the concentrations of the solute and the liquid heights in the branches are both different on each side. This would be a formal argument in addition to the reasons already mentioned for contesting the diagram in

Fig. I.2, which suggests a balance with identical solutions on both sides and different levels in the branches.

It might be thought that once the content of the preceding paragraph has been accepted, the matter is settled. Then common misconceptions on the subject should no longer apply. These include the one already discussed in Chap. 4 – the concentration of solute would be the same on both sides at equilibrium – but this is not the only one (Viennot and Décamp 2016).

Their long-lasting survival among students who have received courses on osmosis prompted us to explore the understanding of the very idea of fluid pressure in populations likely to have difficulty with osmosis.

Water Pressure

It is very likely that common errors on the subject of osmosis are encouraged by a totally inappropriate view of what happens in a liquid when it undergoes a change while remaining liquid. To simplify, we are now talking about water as a prototype solvent.

Some readers may already be wondering what can change in the water if it doesn't change state, in the sense of a phase change (solid, liquid, gas). However, this is what osmosis is all about: the chemical potentials of water in the presence of solute change until they are equal on both sides of the membrane. Osmosis involves physical movements of the water and there are potentially very large differences in pressure.

It may therefore be useful to reconsider the question of what pressure in a liquid is. Let's start with pure water.

Simply put, we know that the temperature and pressure of pure water can change too. But what happens then *in* the water? The point of this question is that it leads us to consider molecular interactions differently than in an academic way, namely as what should be absent in an ideal gas. In fact, these interactions play a decisive role in the pressure within the liquid itself. Let us imagine that we evaluate the pressure of water just below the free surface as for an ideal gas, that is to say, considering that the force necessary to make the molecules that collide with a wall restart in the opposite direction is due exclusively to this wall. Then we would find that this pressure, called "kinetic", is $p_{kin} = NkT/V$ (with p: pressure, V: volume, N number of molecules, k: Boltzmann constant, T: absolute temperature).

But that value would be far too high, since the volume concentration of water, N/V, is about a thousand times higher than that of the gas that is just above the free surface, a gas whose pressure is evaluated by the same formula.

In fact, the predominantly attractive interactions between the molecules of the liquid contribute to "reverse" the movement of the molecules involved in the collisions with the wall, in other words the molecules within the liquid "pull" on those that hit this wall. This is equivalent to negative interaction pressure p_{int}.

This is why the pressure within the liquid is written: $p = p_{kin} - |p_{int}|$.

If the pressure just above the water is the atmospheric pressure, in the water each of the terms of this difference ($p_{kin} - |p_{int}|$) is an order of magnitude a thousand times greater, and it is this difference that is equal to the (atmospheric) pressure of the gas. This shows the considerable importance of molecular interactions in water.

A question asked in interviews of future teachers at the end of their teacher education programmes reveals how new these ideas are to them (Viennot and Décamp *ibid.*). Although they are aware of the existence of these interactions, they do not easily imagine that it is the change in these interactions that is decisive when a glass of water is moved isothermally from sea level to the top of Mont Blanc, where the pressure is half as high. In any case, we must think twice before adopting the commentary by Valentin (1983) on what constitutes a change in the gas-liquid phase.: "(…) molecules have such a low average kinetic energy that they can no longer withstand the electromagnetic attraction they exert on each other" (p. 13). Because, except in the zone where the surface tension occurs (a few molecular diameters below the surface: see Berry 1971; Brown 1974), the pressure in the water remains positive and therefore the kinetic term is higher than that resulting from the attractive interactions.

Table I.1 summarizes elements of questioning and knowledge useful to discuss common documents on osmosis, as well as the level needed a priori to express them, and possibly answer them.

Table I.1 Elements of questioning and knowledge useful to discuss common documents on osmosis

Question to ask ourselves	Knowledge element required[a] and accessible at the end of secondary school	Additional knowledge element required[a] and accessible at university level before graduation
How can a solution be in equilibrium in a U-tube, with, in the other branch, the same solution reaching the same level, but also the same solution reaching a higher level?	Notion of solution	Variation of hydrostatic pressure with altitude. Newtonian equilibrium (second law) in a U-tube.
	Concept of concentration	
	Concept of semi-permeable membrane, symmetry of its properties with respect to both sides.	
	Symmetry with respect to a plane	*Here, the second case (the same solution reaching a higher level) is declared possible, which is contrary to Newton's second law.*
	Here, the second case (the same solution reaching a higher level) is declared possible, which is surprising from the point of view of symmetry.	

(continued)

Table I.1 (continued)

Question to ask ourselves	Knowledge element required[a] and accessible at the end of secondary school	Additional knowledge element required[a] and accessible at university level before graduation
According to the rule stated, what would happen if one branch of the U-tube contained pure solvent and the other a solution?	Notion of solution Concept of concentration Concept of semi-permeable membrane *Here, according to the rule stated, the pure solvent would pass indefinitely to the side of the solution, which is questionable from an energy point of view.*	
What are the variables on which the chemical potential of the solvent in the presence of solute depends?		It is the chemical potential μ of the solvent in the presence of solute that determines the osmotic equilibrium. This quantity must be the same on both sides of the membrane. This quantity, for a given solvent, depends on the absolute temperature T, the pressure p and the molar concentration C. it is not necessary to know the formula holding for dilute solutions of small molecules: $\mu(T,p,c) = \mu_0(T,p)-CkT$ *($\mu_0(T,p)$: chemical potential of the pure solvent, k Boltzmann constant).* *So at equilibrium at given T, either pressure and concentration are identical on both sides of the membrane or they are both different.*
What is pressure in a fluid? Why is it the same in the liquid and its vapor at the thermodynamic equilibrium at the interface, although the liquid phase is about a thousand times denser than the vapor phase?	Pressure in a fluid, related to a normal force exerted on the walls and on a pressure gauge. Pressure due to the impact of molecules on a wall. The effect of these collisions depend on temperature and density. If only the wall acts on the reversal of molecules, then pressure at equilibrium is $P_{kin} = NkT/V$	Intermolecular attractions reduce the effect of wall impacts and introduce a new term - $\lvert p_{int}\rvert$ in the calculation of the equilibrium pressure in the fluid: $p = NkT/V - \lvert p_{int}\rvert$ It is not easy to calculate this term. To give an analytical account of osmosis, it is essential to use the chemical potential of the solvent in the presence of solute.

[a]Required: at least to ask the question, and possibly answer it

References of Appendix I

Berry, M. V. (1971). The molecular mechanism of surface tension. *Physics Education, 6*, 79–84.

Brown, R. C. (1974). The surface tension of liquids. *Contemporary Physics, 15*(4), 301–327.

Diu, B., Roulet, B., Lederer, D., & Guthmann, C. (1989). *Mécanique Statistique.* Paris: Hermann.

Valentin, L. (1983). *L'univers mécanique.* Paris: Hermann.

Viennot, L., & Décamp, N. (2016). Conceptual and critical development in student teachers: First steps towards an integrated comprehension of osmosis. *International Journal of Science Education, 38*(14), 2197–2219.

Appendix J. Bank of Texts That Can Be Used in Sessions of Education to Critique

This appendix presents a bank of texts that can be used as a resource for critical analysis education at different levels. The second column gives all the references to analyses or comments made in this book about them. Finally, the third column proposes a characterisation in terms of 'types of flaws' – reasons to contest an explanation – and/or' risk factors' – or reasons to be vigilant about the foreseeable effect on readers or listeners. Note also the frequent association, in the third column, of the type of fault "logical incompleteness" and the risk factor "a forgotten variable – or phenomenon", a case often related to the expression "abusive generalization". This is in line with the remarks made in Appendix A, echoing Duhem's "logical impossibility" (p. 60): the conclusions drawn from an experiment are always exposed to the risk of ignoring a variable relevant to the phenomenon under study.

Among the risk factors, a 'story-like ', or 'linear causal 'explanation is for us of the 'echo-explanation' type since it takes up a typical structure of common reasoning. To avoid redundant references, the table only mentions the other echo-explanations. In any case, the proposed categorisation can be discussed, the interest is above all to draw the attention of the trainer and/or the trainees to elements requiring clarification.

Theme and possibly 'text' discussed *Type of document*	References in the book	Type of flaws *Type of risk factor*
Two successive statements, validated by an entire class: 1- You can protect yourself from the cold with a survival blanket (debates and manipulations on heat conduction lead to statement 2). 2- With aluminium, you cannot protect yourself from the cold. *College class session, instructional video DVD.*	p. 11 Appendix B	Explicit internal incoherence associated with an abusive generalization: In statement 2, the radiative property of aluminum is ignored. *Only one variable considered and experimentally tested: thermal conduction*
Capillary rise and pressure difference at the air-water interface: '(...) the pressure immediately below the meniscus has a higher value than that of the atmosphere just above. It is this imbalance that explains the rise (...).' Further on: 'The smaller the tube diameter, the greater the meniscus curvature (...): according to Laplace-Young's law, the greater the water depression in the tube.' *Popular text.*	p.12 Appendix F	Explicit internal incoherence: the pressure under the meniscus is successively declared 'higher' and lower ('greater water depression') than that of the atmosphere. *Two linear causal explanations (one cause, one effect): in the first case, it would be a thrust which would explain the ascent, in the second it would be the meniscus which would have drawn the water (all the more so as it is more curved), thus creating a greater depression....*
'"Reaction can no longer balance Action.' *School Book*	p.13	Explicit contradiction of Newton's Third Law *Confusion between forces involved in a force balance (for one system) and those involved in an interaction (between two systems).*
'The force $\vec{F}$ exerted by the cyclist on the bike is directed according to $\vec{v}$ (bike velocity)' *Science book exercise (grade 12)*	p. 14	Indirect contradiction of a law: according to Newton's third law, the cyclist would then be pushed back and ejected from the bike. *Confusion between the person responsible for the movement of the bicycle and the external force acting on the cyclist-bicycle system.*
Radiocarbon dating: The ratio of $^{14}C/^{12}C$ concentrations in the atmosphere is stable over time. *Documents on the Internet*	p. 18 Appendix C	Logical incompleteness: how to explain that the ratio of $^{14}C/^{12}C$ concentrations in the atmosphere is stable over time, when we have just said that radiocarbon disintegrates spontaneously? *The possible intervention of other components in the atmosphere, which generate radiocarbon at the same rate as it decays, is not mentioned.*

(continued)

Theme and possibly 'text' discussed *Type of document*	References in the book	Type of flaws *Type of risk factor*
'In a transparent medium, such as glass for example, light propagates less quickly because its refractive index is higher than that of air'. *Popularisation brochure*	p.17	Logical incompleteness: this is a tautology since the refractive index n of a medium is defined by $n = c/v$, where c and v are the values of the speed of light propagation in vacuum and in the medium respectively.
'When the gas in the water comes into contact with clear limewater, it becomes cloudy. The limewater becomes cloudy in the presence of carbon dioxide. So the gas in the sparkling water is carbon dioxide.' *Current school use*	p.17	Logical incompleteness: the possibility of other causes producing the same effects is not considered. *Only one cause is considered*
Bernoulli's theorem illustrated via a cylindrical stream tube *Popularisation for training purposes*	Appendix D	Indirect contradiction of a law: for an incompressible liquid in steady state, there can be no acceleration in a cylindrical stream tube.
Bodies denser than water do not float. *Current statement*	p. 19	Logical incompleteness and overgeneralization *Only one variable is considered: what determines Archimedes' upthrust (beyond the density of water) is the displaced volume of fluid. This depends not only on the density of the body but also on its shape and the way it is immersed (see the case of boats).*
In a greenhouse, more energy enters than exits. *Common idea, common statement*	p. 20	Incompatibility with a thought experiment: this could not last. *Implicit story-like explanation (we follow the adventure of radiation). No distinction between the transient and the permanent regime, implicit centration on a transient regime.*
"In a gas, collisions between molecules produce heat" *Common idea*	p.20	Incompatibility with a thought experiment: this could not last in a thermally isolated chamber. *Implicit story-like explanation: the permanent regime is not envisaged in the long term.* *Designation of entities: meaning of 'heat'*

(continued)

Theme and possibly 'text' discussed *Type of document*	References in the book	Type of flaws *Type of risk factor*
To save energy, it is better to set the antifreeze protection at 10 °C than 4 °C, taking the cost of heating when you return into account. *Common idea*	p. 20	Incompatibility with a thought experiment; imagine a very long absence during a very long winter: the energy lost to the outside would clearly be greater with 10° inside than with 4°. *Implicit story-like explanation: focus on a transient*
Consider a pierced water bottle resting on a horizontal support. The maximum range, on the support, of a jet from this bottle is obtained with the lowest hole. *Numerous documents*	p. 21 Appendix E	Logical incompleteness Incompatibility with a thought experiment: the limit case of a hole at the level of the support *Only one variable is considered: the waterfall time is disregarded*
A battery in a closed circuit: the field in the wires is due to the charges on the poles of the battery. *Common idea*	p. 23 Appendix F	Indirect contradiction of a law: The wires do not necessarily have the bean shape of the field lines of a dipole. *One aspect of the situation is disregarded: the surface charges on the wire.*
'(…) the free electrons of the circuit are attracted to the positive terminal, repelled by the negative terminal of the battery. They circulate from the negative terminal to the positive terminal outside the generator.' *Common explanation*	p.23	Indirect contradiction of a law: The physical laws seem to be different inside and outside the battery. *In the absence of justification on the potential inside the battery, it is likely to be understood that negative charges, anions, are not subjected, locally, to a force opposite to the electric field that prevails at the point considered.*
'The experiment demonstrating the composition of air by Lavoisier' *A current school presentation*	p. 28	Logical incompleteness *Only one variable is considered: under the bell after oxidation of the mercury, there could be one or several other gas than the nitrogen.* *The accuracy of the conclusion: Lavoisier found an approximately correct result (we know now that other gases are present in the air, in small proportion).*
The isobaric hot air balloon *Common use in school and university exercises*	p. 29, 40	Explicit contradiction of the law of hydrostatics *The accuracy of the conclusion: with this hypothesis and Archimedes' theorem, we find a very acceptable value of the temperature necessary for take-off.* *Small assimilated to zero: pressure variations with altitude are very small but essential to lift.*

(continued)

Theme and possibly 'text' discussed *Type of document*	References in the book	Type of flaws *Type of risk factor*
'From a given drop, only one wavelength reaches the observer's eye'. *University course book*	p.30	*Designation of entities: shortcut and materialization of a concept ('only one wavelength reaches the observer's eye').*
'Pressure is a force distributed over an area' *Popularisation book*	p.31	*Designation of entities: shortcut (A pressure is a force...) involving quantities that do not have the same dimension (same unit).*
'Red pigment absorbs (all the) green light.' *Common use in school books*	p.31	In case we think: 'all the green light': Incompatibility with a thought (or real) experiment *All or nothing. In fact, even with a very low diffusion coefficient, a pigment that receives intense monochromatic radiation diffusively reflects enough light so that the impact zone appears the colour of this light.*
'Liquid water is non-compressible' *Common use in school books*	p. 33	Incompatibility with a thought experiment (if nothing changes in the water at microscopic scale, how does the pressure change?), and with measurements. *Small assimilated to zero:* the coefficient of isothermal compressibility χ_T of water is non zero: $\chi_T = -(1/V)(\Delta V/\Delta p)_T = 4{,}4\ 10^{-10}\ Pa^{-1}$ *In fact 200 m below the water surface, the volume of one liter is 1 cm*^{3} *(order of magnitude) smaller than at sea level.*
'In the absence of atmosphere'... a balloon in the stratosphere *A scientific journal article used for school purposes (grade 10)*	p.33	Explicit contradiction of the law of hydrostatics Incompatibility with a thought experiment: you cannot float without a surrounding fluid. The pressure at 40 km of altitude is very low but essential for lifting. *Small assimilated to zero*
Pascal has demonstrated the existence of a vacuum (by means of a tube turned over a container filled with mercury). *Current statement in school documents*	p. 33	Logical incompleteness: there could be an invisible substance at the top of the tube: Mercury vapour is not mentioned. *Small assimilated to zero*

(continued)

Theme and possibly 'text' discussed *Type of document*	References in the book	Type of flaws *Type of risk factor*
The interest, for a rocket, to take off from Kourou *School book (grade 9)*	p.33	Logical incompleteness *Only one cause is considered: the distance to the centre of the Earth. There could be (and there is) another concomitant cause: the rotation of the Earth.* *Accuracy of factual statements*
How can we explain the absence of atmosphere on certain planets? It is suggested that the low value of *g* is the cause of this phenomenon. *School book (grade 9)*	p. 34	Logical incompleteness *Only one cause is considered: lower value of g. There could be another concomitant cause: (e.g. larger value of the temperature).* *Accuracy of factual statements*
Floating wind turbines: hollow out the platform 'creates an Archimedes' upthrust'. *Radio broadcast*	p. 34	Explicit contradiction of the law of hydrostatics: in fact, as soon as there is submergence, there is an Archimedes' upthrust. *The effect of two causes (factors involved in the flotation: weight, Archimedes' upthrust) is attributed to only one: Archimedes' upthrust, which is nevertheless greater for a platform not hollowed out lying on the bottom of the water than for another of the same external volume, hollowed out, and floating..*
Water-filled test tube turned over on a tank full of water: 'What is lifting this column of water up by 2 m? It's atmospheric pressure that is pushing on the water in the container. In the tube, there is no air, and no pressure is exerted on the water'. *Marie Curie courses for children*	p.35	Indirect contradiction of a law: the atmospheric pressure that presses on the water in the container generates an upward force about five times the weight of the column. *Only one cause envisaged, only one place (at the bottom of the column) whereas there is repulsive interaction between the inside air and the top of the tube.* *Echo explanation: the weight of the water in the column would be equal to the action of this water on its support (the water in the tank), which is false.*
The Earth cannot exert a forward force on the pedestrian that accelerates forward because, at the same time, the support point is moving backwards. *Journal article on physics education*	p. 36	Indirect contradiction of a law: the pedestrian's center of mass accelerates horizontally, therefore the pedestrian at least undergoes an external force with a horizontal component in the direction of acceleration, which only the Earth can exert. *Echo explanation: a force should always be in the direction of movement of its point of application, which is wrong.*

(continued)

Theme and possibly 'text' discussed *Type of document*	References in the book	Type of flaws *Type of risk factor*
Two springs in series hanging from the ceiling, pull down: 'the bottom spring stretches and after a while the second one too'. *Student's response (grade 10) to a survey*	p.39	Indirect contradiction of a law: in quasi-static analysis, the contact interactions at both ends of the springs have the same value. *Explicit story-like explanation (linear causal)*
In a gas: 'fewer molecules' means 'less pressure'. *Current statement, a popularization book*	p. 40	Logical incompleteness and overgeneralisation *Pressure is linked to a single variable, temperature is disregarded. Note that in a hot air balloon there are 'fewer molecules' (than if the air were cold) and there is no 'less pressure'.* *Echo-explanation: pressure seen as a compaction*
'By heating a gas, the pressure is increased', *or* 'Compressing a gas (increasing the pressure) increases the temperature', *or* 'Heating a gas increases its volume'. *Responses from university students during a survey*	p. 40	Logical incompleteness and over generalisation: other variables might affect the value of pressure. *In the triplet pressure, volume, temperature, each variable is linked to only one other and the statements do not specify what happens to the third one.*
Isobaric heating of an ideal gas: 'the temperature increases then the pressure will increase, therefore the volume too'. *Responses from university students during a survey*	p. 40	Explicit internal contradiction (see 'pressure will increase' while heating is said to be 'isobaric') *Explicit story-like explanation (linear causal)*
The siphon 'The water contained in the long branch of the siphon flows out. A vacuum is created and atmospheric pressure causes the water in the container in which it is immersed to rise in the small branch'. *Marie Curie courses for children*	p. 41	Logical incompleteness: there is the same atmospheric pressure value at the free surfaces of the water, in the tank as in the tube open downwards, so why would this pressure cause the water to "rise" and not "fall"? *Explicit story-like explanation (linear causal)* *Echo-explanation: an open branch downwards would necessarily let the water flow.*
'At the summit of Everest (8.848 m), air is becoming thinner: the Earth's attraction is only 9.760 N/kg instead of 9.811 N/kg at sea level'. *School book (grade 9)*	p. 41	Logical incompleteness: air would become thinner even with constant value of "the Earth attraction", on the other hand "the Earth attraction" does not depend on air density. *Echo explanation (suggestion of an improper relationship between density variation and variation of "the Earth attraction")* *Implicit meaning of ':'* *Accuracy of factual assertions*

(continued)

Theme and possibly 'text' discussed *Type of document*	References in the book	Type of flaws *Type of risk factor*
An image to explain the Hall effect: the electrons are deflected and then return in the axis. *Image on Wikipedia in French*	p. 42	Indirect contradiction of a law: In steady state, electrons do not have a velocity component perpendicular to the edge of the sample. *Suggested story-like explanation (linear causal)*
Changes of state by exhaustion 'By decreasing the temperature, the molecules have an average kinetic energy so low that they can no longer resist the electromagnetic attraction they exert on each other: they begin by agglutinating in the liquid state and end up binding in the solid state'. *University course book*	p.44 Appendix I	*Explicit story-like explanation (linear causal)* *Suggestion that:* *At thermodynamic equilibrium (also), the average kinetic energy per particle is less in the liquid than in the gas.* *Hence there is a risk of contradiction of physical laws: In fact, the average kinetic energies at equilibrium between phases at given T are equal.* *Metaphorical language* *Also: If molecules can no longer resist, how can the pressure remain positive?*
The rays seen with a magnifying glass *Image in university course*	p. 47	*Realism v/s Symbolism* *Echo-explanation (light would be visible in profile)*
Sunbeams and lines of sight University course book	p.49	*Similarity of symbols used for different entities* *Echo explanation*
Young's holes that deviate light rays *Image of current school or university use*	P. 50	*Similarity of symbols for different entities (before and after the hole)* *Selection of the elements represented* *Story-like explanation (suggested: the ray continues its adventure)*
Colour syntheses *Commonly used image*	p.51	*Similarity of symbols used for different entities* *Overselectivity of the elements represented* *All or nothing*
Aliens and the size of the Earth *Proposed image for school use (grade 10)*	p.52	Indirect contradiction of a law (what about the size of the Earth on the image whereas it is so far away?) *Image structure and scales of representation*

(continued)

Theme and possibly 'text' discussed *Type of document*	References in the book	Type of flaws *Type of risk factor*
'The cosmic fossil radiation, which the expansion of the universe abandoned on the side of the road about ten billion years ago, and which has since bathed interstellar space without interacting with matter (...) it was discovered by chance in 1964 by two radio astronomers who were looking for something else'. *Popularisation book*	p. 53	Explicit internal contradiction (radiation detected without interacting with matter) *Explicit story-like explanation* *Metaphorical style*
Osmosis leads to an equilibrium with the same solute concentrations and different pressures on either side of the membrane: 'Since everything happens as if the membrane does not exist for the solvent, a situation of imbalance will be created (...) the passage of the solvent taking place until a new equilibrium is established (minimum free energy), therefore when the concentrations in A and B will become equal. There will thus be a difference in level between the compartments (...), therefore a difference in pressure $\Delta p = pA\text{-}pB$, (...) which is called osmotic pressure.' *University course book, similar diagrams on Wikipedia* 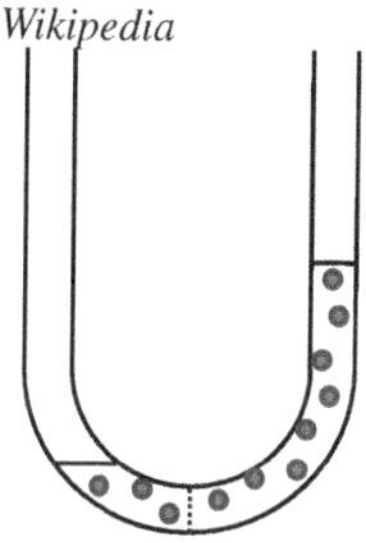 Semi-permeable membrane Final state	p.57, 77 Appendix I	Indirect contradiction of a law Contradiction with two thought experiments (starting with pure water on one side only: one would then have an endless migration of pure water towards the solution; comparing the balance with the same concentrations and pressures on each side of the membane and that described by the text. *Only one cause (only one variable) is considered* *Explicit story-like explanation* *(Iconic aspects could be criticised but are not essential here)*

(continued)

Theme and possibly 'text' discussed *Type of document*	References in the book	Type of flaws *Type of risk factor*
The capillary rise and the angle of contact: the force that 'lifts the liquid' *Common image in text books* 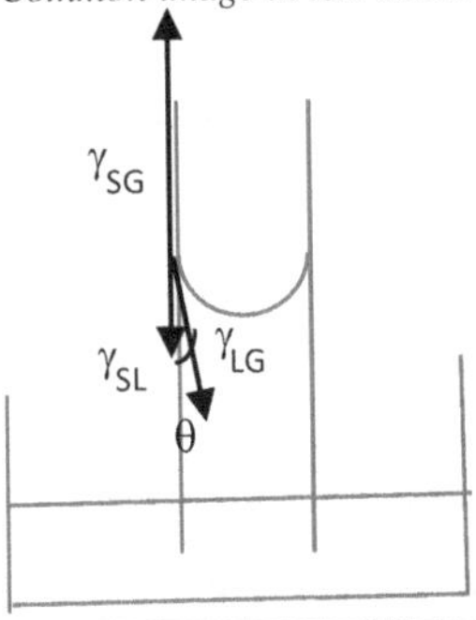	p. 78 Appendix G	Indirect contradiction of a law Logical incompleteness (Balance of horizontal forces? What about the weight of the column?) *Accuracy of conclusion* *A single place (of action of the forces: 'above') is considered* *Overselectivity of the schema (on what objects do the forces act?)* *Metaphorical style (the force "lifts" the liquid)*
To protect yourself from the cold with a survival blanket, you must put the silver side inwards to reflect the heat back towards you. *Common idea* *Instructions for use on products and internet documents*	p. 75 Appendix B Appendix H	Logical incompleteness: what about the radiation of the blanket towards the outside? *Explicit story-like explanation,* *Only one location envisaged (between the body and the cover, and not between the cover and the outside).*
The unfortunate Captain Haddock: the asteroid Adonis attracts him.... *Hergé: We walked on the moon*	p. 85	Indirect contradiction of a law: why would Captain Haddock's trajectory be more deviated than that of the rocket whose engine is off?

The manufacturer's authorised representative in the EU is Springer Nature Customer Service Centre GmbH, Europaplatz 3, 69115 Heidelberg, Germany. If you have any concerns regarding our products, please contact ProductSafety@springernature.com

Printed and bound by CPI Group (UK) Ltd, Croydon, CR0 4YY
22/09/2026
02221058-0019